Student Study Guide

to accompany

Essentials of The Living World

Second Edition

George B. Johnson
Washington University

Jonathan B. Losos
Harvard University

With contributions by
Linda D. Smith-Staton
Pellissippi State Technical Community College

Boston   Burr Ridge, IL   Dubuque, IA   New York   San Francisco   St. Louis
Bangkok   Bogotá   Caracas   Kuala Lumpur   Lisbon   London   Madrid   Mexico City
Milan   Montreal   New Delhi   Santiago   Seoul   Singapore   Sydney   Taipei   Toronto

The **McGraw·Hill** Companies

Student Study Guide to accompany
ESSENTIALS OF THE LIVING WORLD, SECOND EDITION
GEORGE B. JOHNSON AND JONATHAN B. LOSOS

Published by McGraw-Hill Higher Education, an imprint of The McGraw-Hill Companies, Inc., 1221 Avenue of the Americas, New York, NY 10020. Copyright © 2008 by The McGraw-Hill Companies, Inc. All rights reserved.

1 2 3 4 5 6 7 8 9 0 QSR/QSR 0 9 8 7 6

ISBN: 978-0-07-328227-5
MHID: 0-07-328227-8

www.mhhe.com

Contents

Johnson *Essentials* 2e SSG

1   The Science of Biology

Key Concepts Outline

All living things share five fundamental properties. (page 3)
- They are: cellular organization, metabolism, homeostasis, reproduction, and heredity.

There are hierarchies of increasing complexity associated with living things. (pages 4-5)
- The complexity of life can be examined at three levels: cellular, organismal and populational; emergent properties result at each higher level in the living hierarchy.

There are many ways to study biology. (pages 6-7)
- Five general themes often used to organize the study of biology are: the flow of energy, evolution, cooperation, structure determines function, and homeostasis.

The scientific process involves using observations, experimentation, and reasoning (page 8)
- Inductive reasoning is a way to discover general principles by examination of specific cases.

There are many examples of science in action. (page 9)
- The discovery of how CFCs are reducing atmospheric ozone levels is one example.

There are six stages of scientific investigation. (pages 10-11)
- These stages are observing, forming a hypothesis, making predictions, testing the predictions, controlling variables, and forming conclusions.

Once a scientist has conducted his/her research, the results must be analyzed and presented for interpretation by others. (pages 12-13)
- Independent variables are controlled during experimentation and the dependent variable is the response measured and reported by the investigator. Relationships between variables can be reported as correlations or in various kinds of graphs.

Theories are generally accepted scientific principles. (page 14)
- Theories are unifying explanations for many observations; they may be modified as new information is shared throughout the scientific community.

One of the most creative aspects of scientific investigation is the formulation of novel hypotheses. (page 15)
- Scientists have a sense of imagination.

There are four unifying theories of biology. (pages 16–20)
- These are the cell theory, the gene theory, the theory of heredity, and the theory of evolution.

Key Terms Matching

1. _____ Properties
2. _____ Levels of organization
3. _____ Themes
4. _____ Inductive reasoning
5. _____ Hypothesis
6. _____ Controlled experiment
7. _____ Theory
8. _____ Scientific method

a. A way to answer questions.
b. A proposed explanation to a question or problem.
c. Discovering general principles by careful examination of specific cases.
d. All living things respond to stimulation.
e. Atoms, molecules, cells, etc.
f. Only one variable is changed at a time.
g. A hypothesis that has withstood the test of time.
h. Structure determines function, the flow of energy.

1.1   The Diversity of Life (Page 2)

9. The diversity of life is separated into _____ kingdoms.
10. Methanogenic bacteria are members of kingdom _____.
11. Kingdom _____ contains the multicellular photosynthetic organisms.

1.2 Properties of Life (Page 3)

12. A set of instructions that determines the characteristics of an organism is called a(n)
    a. membrane.   b. cell.   c. nucleus.   d. gene.   e. atom.

13. A _____ is a tiny compartment covered by a thin membrane. It is the basic unit of life.
    a. nucleus   b. mitochondrion   c. cell   d. tubule   e. tissue

14. Which of the following is a property of life?
    a. Response to stimulation   b. Homeostasis   c. Metabolism
    d. Reproduction   e. All of the above

15. _____ is the use of energy by living things.

16. The genetic systems seen in all living organisms are based on a molecule called _____. This molecule is organized into units called _____.

1.3 The Organization of Life (Pages 4–5)

17. _____ Organs   a. Biological machines comprised of various tissues.
18. _____ Organelle   b. Cellular activities occur at this level.
19. _____ Species   c. A group of interbreeding organisms.

20. _____ are made of cells sharing similar structure and function.
21. An ecosystem is made up of _____ and _____.

1.4 Biological Themes (Pages 6–7)

22. _____ Structure determines function   a. Water balance is an important part of this process.
23. _____ Evolution   b. The long tongue of a moth allows the insect to reach nectar.
24. _____ Homeostasis   c. Natural selection is the driving force in this process.

25. _____ wrote the book _____ in which he discussed his theory of evolution by natural selection.

1.5 How Scientists Think (Page 8)

26. Computers work by use of _____ reasoning.
27. Discovering general principles by examining specific cases is called_____ reasoning.

1.6 Science in Action: A Case Study (Page 9)

28. _____ atmospheric ozone layer   a. Depletion of the ozone layer.
29. _____ CFCs   b. Protects life from harmful UV rays.
30. _____ Increased incidence of skin cancer.   c. Destroy ozone.

1.7 Stages of a Scientific Investigation (Pages 10–13)

31. The test of a hypothesis is a(n)
    a. experiment.   b. challenge.   c. new hypothesis.   d. theory.   e. question.

32. The key to a successful scientific investigation is
    a. developing a correct hypothesis.   d. including two or more controls in the experiment.
    b. developing a variety of hypotheses.   e. making careful observations.
    c. a complex experiment.

33. A factor that might influence a process is called a(n)
    a. control. b. variable. c. question. d. conclusion. e. prediction.

1.8   Theory and Certainty (Page 14)

34. A series of steps used to answer scientific questions is called the _____.
35. _____ questions cannot be answered by scientists.

1.9   Four Theories Unify Biology as a Science (Pages 16–18)

36. A _____ consists of the entire set of DNA instructions.
37. Genes are located on _____.

Chapter Test

1. The science of biology
   a. explores the magical processes by which our world operates.
   b. is conducted by people with poor organizational skills.
   c. discovers the general principles of our world by applying inductive reasoning.
   d. explores a variety of problems using the "gut" instincts of scientists.
   e. both a and d.

2. Using an algebraic formula to calculate the load capacity of a bridge is an example of
   a. deductive reasoning.   d. applied math.
   b. inductive reasoning.   e. mathematical reasoning.
   c. theoretical math.

3. Most new cars are manufactured with air conditioners that no longer use chlorofluorocarbons (CFCs) as coolants. This is in part due to
   a. their role in the reduction of light intensity in the atmosphere.
   b. their presence in the upper atmosphere for over 100 years.
   c. the allergic reaction that occurs in most people when they inhale CFCs.
   d. their role in the destruction of the ozone layer.
   e. both b and d.

4. Which of the following are in the correct order?
   a. Test, make observations, develop a hypothesis, make predictions, draw conclusions.
   b. Make observations, develop a hypothesis, make predictions, test, draw conclusions.
   c. Develop a hypothesis, test, make predictions, make observations, draw conclusions.
   d. Make predictions, develop a hypothesis, test, make observations, draw conclusions.
   e. either a or b.

5. In an effort to improve crop yield a rancher tries different kinds of fertilizers. She uses an organic fertilizer on one field and a chemical fertilizer on another. A third field is not fertilized. The two fertilizers represent what part of her experiment?
   a. The hypothesis   b. The control   c. Variables   d. Predictions   e. A theory

6. The unfertilized field in question 5 represents what part of her experiment?
   a. The hypothesis   b. The control   c. Variables   d. Predictions   e. A theory

7. The scientific method is limited to
    a. events that occur in our solar system.
    b. supernatural phenomena.
    c. religious questions.
    d. mathematical problems.
    e. observable and measurable processes.

8. A group of individuals of the same species living together is referred to as a(n)
    a. population.   b. community.   c. species.   d. ecosystem.   e. biome.

9. A tropical forest is an example of a(n)
    a. population.   b. community.   c. species.   d. ecosystem.   e . biome.

10. While walking through a deciduous forest in New England one might observe squirrels, blue jays, gypsy moths, sugar maples, and blue spruce. All of those organisms represent a(n)
    a. population.   b. community.   c. species.   d. ecosystem.   e. biome.

11. A bacterial colony consists of identical cells that are derived from a single original cell. Those cells represent members of the same
    a. kingdom.   b. community.   c. species.   d. ecosystem.   e. biome.

12. Over 300 breeds of purebred dogs exist today. They are the result of
    a. evolution.   d. selective selection.
    b. natural selection.   e. adaptive evolution.
    c. artificial selection.

13. Scientists believe that the ozone layer over northern Europe may begin to break down because
    a. they have detected elevated levels of chlorine in the upper atmosphere in that area.
    b. the number of cases of skin cancer in that area has increased in recent years.
    c. crop yields are down in the area due to increased levels of ultraviolet radiation.
    d. that is the next likely place to be affected after Antarctica.
    e. none of the above.

14. Ultraviolet light causes skin cancer by
    a. damaging the organelles in skin cells resulting in cell death.
    b. damaging the DNA in skin cells.
    c. increasing the quantity of melanin in skin cells.
    d. concentrating toxins in skin cells.
    e. both a and c.

Additional Study Help

Visit the ARIS (Assessment, Review, and Instruction System) site at aris.mhhe.com for quizzes, animations, and other study tools.

2   Evolution and Ecology

Key Concepts Outline

In 1831 Darwin sailed around the world. (pages 24–25)
- He closely observed the plants and animals he saw. Darwin was able to explore the biological richness of the tropical forests of South America.

Darwin sought to explain his observations of fossils of extinct organisms similar to currently living organisms and the Galapagos finches. (page 27)
- Darwin believed the fossilized organism must have given rise to the currently living organisms and that the finches adapted to different sources of food resulting in many variations in beak shape and size.

Darwin's greatest contribution to science is his formulation of the hypothesis that evolution occurs as a result of natural selection. (pages 27-28)
- Individuals with characteristics more suitable for survival and reproduction will tend to leave more offspring, and so become more common in future generations.

Darwin's Galapagos finches were a key component of his argument for evolution by natural selection. (pages 29-30)
- The correspondence between the beaks of different finch species and the available sources of food suggested to Darwin that evolution caused the difference in beak shape. The differences in beaks are the result of differences in the genes of the various finch species and changes in beak shapes are observed even today.

Evolution can result in similar yet unrelated species in various environments. (page 31)
- DNA analysis of Caribbean lizards supports the hypothesis that species may evolve independently to be similar to each other.

Natural selection produces diverse species that occupy different niches in an environment. (page 32)
- Adaptive radiation of an ancestral population results in different species that occupy a variety of different habitats within a region; it is especially common on island groups or in other discontinuous habitats.

Ecology is the study of how organisms interact with each other and the environment in which they live. (page 33)
- There are five levels of organization considered when studying ecology: populations, communities, ecosystems, biomes, and the biosphere.

The most complex biological system studied by biologists is an ecosystem. (page 34)
- Energy flows through ecosystems when it is lost as heat in the steps up a food chain; materials like carbon, nitrogen and phosphorus cycle within ecosystems as organisms perform photosynthesis, respiration, and decomposition.

All the organisms that live together in an area make up a community, the living component of an ecosystem. (page 35)
- A community is comprised of the plants, animals and microorganisms that live together in an area.

When a resource is limited in a community and two organisms attempt to use that resource, competition results. (page 36)
- Interspecific competition occurs between members of different species; intraspecific competition occurs between individuals of a single species.

The principle of competitive exclusion says no two species can coexist within the same niche. (page 37)
- Resource partitioning enables species that live in the same geographical area to avoid competition. Niche overlap is reduced when character displacements evolve in two species.

The consumption of one organism by another is predation. (page 38)
- Coevolution causes reciprocal evolutionary adjustments between prey and predators.

Symbiosis occurs when two or more kinds of organisms interact in fairly permanent relationships. (page 39)
- Commensalism, mutualism and parasitism are different forms of symbiosis.

Key Terms Matching

1. _____ Evolution
2. _____ Natural selection
3. _____ Galápagos Islands
4. _____ *On the Origin of Species*
5. _____ Adaptive radiation
6. _____ Ecology
7. _____ Population
8. _____ Ecosystem
9. _____ Resource partitioning
10. _____ Symbiosis

a. Many niches in an environment lead to this.
b. A long-term interaction between two or more species.
c. How organisms live in their environment.
d. Reduces competition between species living in the same area.
e. Individuals of the same species living together.
f. Darwin's journey led him to develop a theory about this.
g. It challenged all that people believed at that time in history.
h. It regulates the flow of energy. A community and nonliving factors.
i. Many species of finches evolved here and were observed by Darwin.
j. Measured by reproductive success.

2.1 Darwin's Voyage on HMS *Beagle* (Pages 24-25)
True (T) or False (F) Questions
If you believe the statement to be false, rewrite the statement as a true one.

11. Prior to the publication of Darwin's *On the Origin of Species* most people believed that species were created and unchanging over time.
    Answer: _____ Restatement: _____
    _____

12. Darwin's God created species and left them forever unchanged.
    Answer: _____ Restatement: _____
    _____

13. Darwin was one of many scientists who proposed natural selection as the mechanism of evolution.
    Answer: _____ Restatement: _____
    _____

2.2 Darwin's Evidence (Page 26)

14. Fossil evidence studied by Darwin led him to believe in evolution because
    a. it was obvious that simpler but similar organisms must have given rise to modern species.
    b. God would not have left behind fossils if He did not want Darwin to discover evolution.
    c. the fossils were exactly the same as extant species.
    d. the fossil record was so complete.
    e. so many other scientists believed in evolution as evidenced by the fossil record.

15. The writings of Charles Lyell had a great influence on Darwin because Lyell
    a. wrote about natural selection as the mechanism of evolution in great detail.
    b. believed species changed when God brought forth a great disaster and created new forms of life on the renewed Earth.
    c. described a world where species were constantly becoming extinct while new species evolved.
    d. described several species of finches in which different types of beaks allowed them to occupy different niches.
    e. felt that humans had no influence on the evolution of other species.

16. Which of the following helped convince Darwin about the evolution of species?
    a. The fossil record
    b. Patterns of life he observed on the voyage of the *Beagle*
    c. Biblical writings
    d. All of the above helped to convince Darwin about the evolution of species.
    e. Only a and b helped to convince Darwin about the evolution of species.

2.3   The Theory of Natural Selection (Pages 27-28)

17. _____ The Descent of Man
18. _____ Wallace
19. _____ Darwin
20. _____ Hardy-Weinberg
21. _____ Malthus

a. Developed the theory of evolution by natural selection.
b. Wrote essay on the "Principles of Population."
c. Darwin's second book – humans and apes have common ancestors.
d. Wrote a paper on evolution but did not provide as much evidence as was in *On the Origin of Species.*
e. Studied how changes in gene frequencies lead to evolution.

2.4   The Beaks of Darwin's Finches (Pages 29-30)

22. What was especially profound about the different beak types Darwin observed in his "finches" was that
    a. beak type affected the number of offspring produced by each bird.
    b. each beak type was seen on only one island.
    c. the beak type changed over the life time of each bird.
    d. this incredible diversity occurred in a closely related group of birds.
    e. no other life form Darwin found on his journey showed such striking resemblance to mainland organisms.

23. The beak shape of Darwin's finches was influenced by the
    a. number of offspring produced in a given year.
    b. carrying capacity of the environment.
    c. response to the nature of the food supply.
    d. effects of artificial selection by the island's inhabitants.
    e. predators preying on a particular species of bird.

24. Beaks of the finches studied by the Grants changed in response to
    a. the island on which the individual bird resides.
    b. the amount of rainfall each year which in turn influences the type of seeds available.
    c. clutch size produced by a pair of birds which varies significantly each season.
    d. gender of the bird.
    e. both a and c.

2.5   How Natural Selection Produces Diversity (Page 32)

25. Clusters of species evolved relatively recently from a common ancestor are an example of
    a. adaptive radiation.   b. diverse radiation.   c. ecological selection.
    d. artificial selection.   e. niche selection.

26. Diverse selective pressures on a group of organisms often lead to
    a. extinction.
    b. rapid speciation.
    c. the ability to interbreed with another closely related species.
    d. a splinter group moving to a new area.
    e. both b and c.

2.6   What Is Ecology? (Page 33)

27. _____ Population
28. _____ Community
29. _____ Ecosystem
30. _____ Biome

a. Members of the same species living together and sharing resources.
b. This regulates the cycling of essential elements.
c. Different species living together in the same area.
d. Marine and freshwater habitats are examples.

2.7   A Closer Look at Ecosystems (Page 34)

31. The basis of the food chain is/are
    a. the carnivores.   b. the sun.   c. the algae.   d. higher plants.   e. the decomposers.

32. Materials in an ecosystem are
    a. recycled two or three times before they are exhausted.
    b. available only to photosynthetic organisms.
    c. passed back into the ecosystem when an organism dies.
    d. limited to carbon compounds.
    e. lost as heat when an organism dies.

2.8   Communities (Page 35)

33. Which of the following is an example of a community?
    a. redwood trees.
    b. redwood trees, redwood sorrel, sword ferns, slugs and ground beetles.
    c. loamy soil.
    d. head lice and humans.
    e. desert.

2.9   The Niche and Competition (Page 36)

34. Competition between deer and horses for the same grass is _____ competition.

35. Which of the following would involve intraspecific competition?
    a. zebras and wildebeests.
    b. two maple tree seedlings.
    c. rabbits and foxes.
    d. rumen bacteria that digest cellulose for cows.

2.10   How Species Evolve to Occupy Different Niches Within an Ecosystem (Page 37)

36. A place is a(n) _____ while a pattern of living is a (n) _____.
37. Changes that evolve in two species that reduce niche overlap are called _____.

2.11   Predation (Page 38)

38. True/False - Giraffes that eat the acacia tree leaves are predators. _____
39. Defense mechanisms of a prey organism that enable it to escape predation are the result of _____.

2.12   Symbiosis (Page 39)

40. The relationship between a cow and the bacteria that live in its rumen and digest cellulose for the cow is an example of
_____.
41. A cat or dog with tapeworms is an example of _____.

Chapter Test

1. Darwin developed his theories on evolution by natural selection after observing
    a. the feeding adaptations of the Galápagos finches.
    b. fossils of extinct and living armadillos in South America.
    c. similarities and differences in the Galápagos finches and South American finches.
    d. differences in the reproductive success in a variety of organisms.
    e. all of the above.

2. A key point made by Thomas Malthus in his *Essay on the Principles of Population* was that
    a. natural selection is the mechanism of evolution.
    b. natural selection is more significant in wild animals than in domestic animals.
    c. the carrying capacity of any population is limited by the available natural resources.
    d. populations of species remain more or less stable in size.
    e. reproductive success is high in live-bearing species.

3. Evolution
    a. has occurred only in the past 150 million years.
    b. involves a progression of changes that took place over long periods of time.
    c. occurs more often in plant species than in animals.
    d. is no longer possible.
    e. always results in species adapted for the current environment.

4. On his journey aboard the *Beagle* Darwin noted that parts of the world having similar climates, such as California, Chile, South Africa and Australia,
    a. exhibited remarkably similar plant and animal life, illustrating the central role of climate over evolution.
    b. had unrelated plants and animals, illustrating that diversity is not entirely influenced by climate.
    c. showed that human cultures could develop in almost identical ways in different parts of the world.
    d. Both a and c are true.
    e. None of the above are true.

5. The variations in Galápagos tortoises Darwin saw could be distinguished based on
    a. differences in the structures of their shells.
    b. modifications of their beaks which allowed them to take advantage of different foods.
    c. distinctive markings on the shells of their eggs.
    d. the length of time their eggs required incubation.
    e. their size.

6. Darwin found that plant and animal species living on oceanic islands were most similar to species found
    a. at the same latitude anywhere else in the world.
    b. at the same elevation anywhere else in the world.
    c. anywhere in the world having a similar climate.
    d. on the nearest mainland.
    e. on neighboring islands.

7. In his discussion of artificial selection Darwin noted that
    a. species which had developed as a result of natural selection were hardier than those influenced by humans.
    b. species which had developed as a result of natural selection showed more diversity than those influenced by humans.
    c. domesticated species showed more diversity than wild species.
    d. evolutionary change could not occur through artificial selection.
    e. in all cases, artificially selected species were superior to wild species.

8. As energy moves through the food chain
    a. much of it is lost to the next level.
    b. it is concentrated and more is available to the next level.
    c. the animal levels are more efficient in using energy than are plants.
    d. it is efficiently recycled.
    e. new levels are added to the food chain on a regular basis.

9. Raw materials such as carbon and nitrogen
    a. can only be used by plants in their raw forms.
    b. move through the food chain in a manner similar to energy.
    c. are so plentiful that recycling is not necessary.
    d. are more plentiful in some biomes than in others.
    e. cycle between organisms and the physical environment.

10. The competitive exclusion principle says
   a. the same ecological niche can only be occupied by one species.
   b. prey and predator populations cycle in abundance.
   c. both species living together benefit from each other.
   d. energy is lost as heat at each step of the food chain.
   e. materials are recycled when decomposers break down discarded and/or dead tissues.

11. Between which of the following will the competition be most acute?
   a. foxes and rabbits.
   b. robins and blackbirds.
   c. two monarch butterflies.
   d. cows and grass.
   e. maple and pine trees.

12. On Galápagos islands where different finch species are found living together, the two species have beaks of different sizes in order to eat different kinds of seeds. This is an example of
   a. intraspecific competition.
   b. commensalisms.
   c. artificial selection.
   d. character displacement.
   e. warning coloration.

13. Ants that live in the nodes of an acacia tree protect the tree from predatory giraffes and in return receive ant larvae food that the tree produces just for them. This is an example of
   a. parasitism.
   b. commensalism.
   c. mutualism.

Additional Study Help

Visit the ARIS (Assessment, Review, and Instruction System) site at aris.mhhe.com for quizzes, animations, and other study tools.

3   The Chemistry of Life

Key Concepts Outline

All matter is composed of atoms. (pages 44-45)
- Atoms are made up of protons, neutrons, and electrons. An atom is typically described by the number of protons in its nucleus; electrons determine the chemical behavior of atoms.

An atom may gain or lose electrons from its outer shell; neutrons in an atom can vary without changing the chemical property of the element. (pages 46-47)
- An atom that loses an electron becomes a positively charged ion called a cation and an atom that gains an electron becomes a negatively charged ion called an anion.
- Atoms that have the same number of protons, but different numbers of neutrons are called isotopes; radioactive isotopes are used in dating fossils and in medicine.

Molecules are collections of atoms. (pages 48-49)
- Molecules form when atoms are held together by ionic or covalent bonds; hydrogen bonds form when a polar covalent bond forms between atoms.

Water's unique properties and importance to living organisms are the result of hydrogen bonds. (pages 51-52)
- Water is a polar molecule and can form hydrogen bonds that are responsible for the important physical properties of water: heat storage, ice formation, high heat of vaporization, cohesion and high polarity.

Water's covalent bonds sometimes break and form a hydrogen ion and hydroxide ion. (pages 53-55)
- The pH scale is a convenient way to express the hydrogen ion concentration of a solution; an acid increases the concentration of hydrogen ions while a base decreases the number of hydrogen ions.
- Buffers take up or release hydrogen ions into solution as the hydrogen ion concentration of the solution changes, which helps maintain a constant pH.

Key Terms Matching

1.	_____ Atom	a. Repels water.
2.	_____ Electron	b. Dissociates in water and increases the concentration of H^+.
3.	_____ Ion	c. Slows changes of pH.
4.	_____ Chemical bond	d. This type of bond is responsible for the unique characteristics of water.
5.	_____ Hydrogen bond	e. Made of protons, neutrons and electrons.
6.	_____ Polar molecule	f. A measurement of H^+ vs. OH^- in water.
7.	_____ Hydrophobic	g. An atom that has gained or lost one or more electrons.
8.	_____ pH	h. These hold atoms into molecules.
9.	_____ Acid	i. Circles the nucleus of an atom and has a negative charge.
10.	_____ Buffer	j. The nucleus of one atom is much larger than the nuclei of other atoms in the molecule.

3.1   Atoms (Pages 44-45)

11. _____ is/are the smallest particles into which a substance can be divided.
     a. Electrons   b. Atoms   c. Isotopes   d. Neutrons   e. Protons

12. Electrons circle the nucleus of an atom in what are called
     a. spheres.   b. domains.   c. domiciles.   d. orbitals.   e. regions.

13. The maximum number of electrons that the lowest energy level may contain (closest to the nucleus of the atom) is
     a. 2   b. 4   c. 6   d. 8   e. 9

14. Which of the following has the least amount of mass?
     a. A proton   b. A neutron   c. An electron   d. An atom   e. A molecule

15. When an electron loses energy, it
    a. flies away from the atom.
    b. donates that energy for the bonding of atoms into molecules.
    c. moves up to the next energy level (one further away from the nucleus).
    d. moves down to the next energy level (one closer to the nucleus).
    e. loses the potential energy it once contained.

16. An atom with an incompletely filled outermost energy level is likely to be
    a. unable to form bonds with other atoms.   d. an isotope.
    b. reactive.   e. bound to carbon.
    c. extremely stable.

3.2   Ions and Isotopes (Pages 46-47)

17. Which of the following is an anion?
    a. I^- b. Ca^{2+} c. Na^+ d. He   e. H_2O

18. Carbon 12, carbon 13, and carbon 14 have different numbers of _____.
    a. protons   b. neutrons   c. electrons

19. Radioactive decay is the result of:
    a. the breakdown of the nucleus of an unstable isotope of an atom.
    b. electron loss from the outermost energy level of an atom.
    c. ionization of calcium-containing substances.
    d. ozone depletion from the atomsphere.
    e. both b and c.

3.3   Molecules (Pages 48-49)

20. Two characteristics of ionic bonds that make them form crystals are that they are _____ and that they are *not* _____.
21. A characteristic that covalent bonds share with ionic bonds is that covalent bonds are also _____ but are _____.
22. Molecules having slightly charged areas are in a manner of speaking "molecular magnets." These molecules are said to be _____ in nature.
23. Because hydrogen bonds are weak they are also highly _____.

3.4   Hydrogen Bonds Give Water Unique Properties (Pages 51-52)

24. Because they are highly polar, water molecules are _____ to most ions.
25. The attraction of like molecules is _____, while the attraction of different substances is _____.
26. Nonpolar molecules aren't water soluble and are called _____ because they appear to shrink from contact with water.

3.5   Water Ionizes (Pages 53-55)

27. Acidic substances have pH values _____ 7 while basic substances have pH values _____ 7 on the pH scale.
28. In water having a pH of 7 there are present equal quantities of the ions _____ and _____.
29. A substance that acts as a reservoir for hydrogen ions is called a(n) _____.

Chapter Test

1. _____ are used by scientists to determine the age of fossils.
    a. Ions   b. Electrons   c. Isomers   d. Isotopes   e. Acids

2. An atom with eight electrons and six protons is an example of a(n)
   a. molecule.   b. mole.   c. radioactive isotope.   d. ion.   e. polar element.

3. Chemical energy can be stored in the form of
   a. extra electrons in the outermost orbital.
   b. extra protons in the nucleus of an atom.
   c. electrons in "high-energy" orbits.
   d. electrons released from high-energy orbitals.
   e. both a and c.

4. When electrons are not shared equally by atoms a(n) _____ forms.
   a. ion   b. polar covalent bond   c. triple covalent bond   d. ionic bond

5. _____ are the type of bond found in crystals such as table salt.
   a. Ionic bonds   b. Covalent bonds   c. Hydrogen bonds   d. Van der Waals
   e. Both a and d

6. It takes quite a lot of energy to bring water to a boil. This is due to
   a. the covalent bonds in a molecule of water.
   b. the hydrogen bonds between adjacent water molecules.
   c. the ionic bonds that are broken when water reaches the boiling point.
   d. excessive quantities of hydrogen ions found in impure water.
   e. both a and d.

7. If an atom has four electrons in its second energy level, which of the following must be true?
   a. The third energy level must have at least two electrons.
   b. The third energy level must be empty.
   c. The first energy level must have two electrons.
   d. The first energy level must have four electrons.
   e. Both b and c are correct.

8. A research chemist discovers a new element. She determines that the atoms of this element are particularly reactive. That means that the atom probably
   a. has an incompletely filled outer electron orbital.   d. has a high specific heat.
   b. has a filled outer electron orbital.   e. has probably been discovered before.
   c. is radioactive.

9. Ionic bonding involves
   a. the sharing of electrons between two or more atoms.
   b. weak bonding between adjacent salt molecules.
   c. weak bonding between adjacent water molecules.
   d. the attraction of oppositely charged atoms.
   e. both a and d.

10. One reason that ionic bonds do not play an important role in most biological molecules is that they are
    a. incompatible with water.   d. too strong to be broken in the cell.
    b. directional.   e. unable to form in carbon-based molecules.
    c. not directional.

11. Hydrogen bonding in water is a consequence of
    a. the polar covalent nature of water molecules.
    b. the nonpolar bonding that occurs between oxygen and hydrogen.
    c. excessive hydrogen dissolved in the water.
    d. both a and b.
    e. both a and c.

12. Hydrogen bonds are
    a. weak bonds that last only a fraction of a second.
    b. weak bonds that last indefinitely.
    c. weak bonds that hold molecules together.
    d. strong bonds that form between adjacent molecules.
    e. strong bonds that last for only a fraction of a second.

13. Sweating helps runners in a marathon to cool off. This is because
    a. the water molecules in sweat adhere to the skin and have a cooling effect.
    b. for every gram of water that evaporates from their bodies, 586 calories of heat are removed.
    c. the water lost in sweating reduces body weight, making it easier for them to run.
    d. the salt in sweat has a cooling effect on the skin.
    e. both a and d.

14. A common medium used in the microbiology laboratory is carbohydrate fermentation broth. When introduced to the broth, a bacterium that is capable of fermenting the carbohydrate produces acidic waste products. What could the microbiologist add to the medium to slow down changes in its pH?
    a. additional acid   d. a buffer
    b. additional carbohydrate   e. a substance that is a strong base
    c. an additional source of nutrients

Additional Study Help

Visit the ARIS (Assessment, Review, and Instruction System) site at aris.mhhe.com for quizzes, animations, and other study tools.

4   Molecules of Life

Key Concepts Outline

Living organisms form organic molecules consisting of a carbon-based core with special groups attached. (pages 60-61)
- Four macromolecules—proteins, nucleic acids, carbohydrates and lipids—are the building materials of cells.
- Macromolecules are assembled when dehydration synthesis creates bonds between subunits called monomers; hydrolysis breaks up polymers by breaking the bonds between subunits.

Proteins have many functions in the bodies of organisms. (pages 62–66)
- Enzymes facilitate chemical reactions. Some proteins are structural in nature. Others act as chemical messengers.
- Amino acids are linked by peptide bonds to form proteins.
- Four levels of protein structure influence a protein's shape, which determines a protein's function.

The information storage molecules of cells are long polymers called nucleic acids. (pages 67-69)
- DNA and RNA are the two kinds of nucleic acids formed when nucleotides are linked together.

Polymers called carbohydrates make up cells' structural framework and are involved in energy storage. (pages 70-71)
- Simple sugars called monosaccharides can be linked together to form complex carbohydrates called polysaccharides.

Lipids are polymers used for long-term energy storage, as components of the cell membrane, and as signaling molecules. (pages 72-73)
- Lipids are nonpolar; fats are formed by linking fatty acid and glycerol subunits.

Key Terms Matching

1. _____ Macromolecule   a. Arranged into units called genes. The molecule of heredity.
2. _____ Functional group   b. There are two types of this macromolecule: DNA and RNA.
3. _____ Organic molecule   c. These molecules have a ratio of 2:1, hydrogen to carbon atoms.
4. _____ Enzyme   d. Means by which a polymer is broken up.
5. _____ Hydrolysis   e. A collection of amino acids assembled in a very specific manner.
6. _____ Carbohydrate   f. Facilitate molecule positioning so the correct bonds are broken.
7. _____ Lipid   g. Groups of atoms with special properties.
8. _____ Protein   h. Cholesterol is an example of this type of macromolecule.
9. _____ Nucleic acid   i. Molecule with a carbon-based core and attached special groups.
10. _____ DNA   j. Many small, repeating subunits bonded together.

4.1   Polymers Are Built of Monomers (Pages 60-61)

11. When _____ occurs a covalent bond forms to connect two subunits; water is also made.
12. The amino functional group is found in _____.
13. Monosaccharides are the _____ of starch which is a polysaccharide.

4.2   Proteins (Pages 62–66)

14. _____ is a _____ protein that forms the feathers of birds and hair.
15. Enzymes are extremely important participants in cellular reactions because they _____ of chemical reactions.
16. A protein is a _____ composed of _____.
17. The _____ structure of a protein is represented by two or more polypeptide chains bonded together.
18. _____ proteins help new proteins fold correctly.

4.3 Nucleic Acids (Pages 67-69)

19. The three parts of a nucleotide are _____, _____, and _____.
20. In DNA adenine pairs with _____ and cytosine pairs with _____.
21. The two strands of a DNA molecule are bonded together by _____.
22. The "backbone" of DNA/RNA is composed of _____ and _____.

4.4 Carbohydrates (Pages 70-71)

23. Long chains of sugars are called _____.
24. The disaccharide sucrose forms when _____ links a glucose and a fructose.
25. Complex carbohydrates are excellent molecules for energy storage because they have many _____.
26. A structural polysaccharide common to fungi and insects is _____.
27. Animals store energy in the form of _____ while plants store _____.

4.5 Lipids (Pages 72-73)

Matching

28. _____ Phospholipid   a. Composed of four carbon rings.
29. _____ Saturated   b. This molecule has a polar group at one end and two long nonpolar tails.
30. _____ Steroid   c. Fatty acid with the maximum number of hydrogen atoms; solid at room temperature.

Chapter Test

1. Hydrolysis of starch yields
   a. amino acids.   d. glycerol.
   b. nucleotides.   e. fatty acids.
   c. monosaccharides.

2. Animal cell membranes contain the steroid
   a. glucose.   b. starch.   c. chitin.   d. cholesterol.   e. RNA.

3. Energy is carried to your body cells by the monosaccharide
   a. glucose.   b. starch.   c. chitin.   d. cholesterol.   e. RNA.

4. Dehydration synthesis forms a _____ that links two amino acids.
   a. hydrogen bond   d. ion
   b. isotope   e. peptide bond
   c. double covalent bond

5. How do DNA and RNA differ?
   a. They have different ribose sugars.
   b. DNA is a double-stranded molecule while RNA is single-stranded.
   c. Thymine is present in DNA but not in RNA.
   d. Uracil is present in RNA but not in DNA.
   e. All of the above.

6. Hydrolysis of a nucleic acid produces
   a. carbohydrates.   d. nucleotides.
   b. fatty acids.   e. steroids.
   c. amino acids.

7. All other levels of protein structure are determined by a protein's _____ structure, the sequence of its amino acids.
   a. primary   b. secondary   c. tertiary   d. quaternary

8. The separation of oil and vinegar in a bottle of Italian salad dressing occurs because the oil is
   a. polar.   d. saturated.
   b. hydrophobic.   e. an ion.
   c. composed of 4 carbon rings.

9. The sequence of _____ in a molecule of DNA determines the sequence of amino acids in a protein.
   a. fatty acids   d. monosaccharides
   b. nucleotides   e. sugar and phosphate groups
   c. peptides

10. All of the following are subunits of a polymer except a(n)
    a. amino acid.   d. monosaccharide.
    b. DNA.   e. fatty acid.
    c. glycerol.

Additional Study Help

Visit the ARIS (Assessment, Review, and Instruction System) site at aris.mhhe.com for quizzes, animations, and other study tools.

5   Cells

Key Concepts Outline

The cell theory says all living organisms are composed of cells. (pages 78-81)
- Cell size is limited by the distance to which substances must diffuse through the cytoplasm and also by the surface-to-volume ratio as cell size increases.
- Biologists are able to see cells using different types of microscopes and staining techniques.

The plasma membrane encases all living cells. (pages 82-84)
- The fluid mosaic model describes the plasma membrane as a phospholipid bilayer with proteins embedded in it.
- The proteins identify cells, bind certain messenger chemicals and act as passageways for chemicals to enter and leave the cell.

Prokaryotes are the simplest cellular organisms and are either bacteria or archaea. (page 85)
- The key characteristic shared by all prokaryotes is that they lack internal compartments.

Eukaryotic cells are internally compartmentalized. (pages 86-87 and 90-92)
- The endomembrane system creates the internal compartmentalization of eukaryotic cells.

The nucleus is the control center of a eukaryotic cell. (pages 88-89)
- DNA, the hereditary information, is located in the nucleus in a eukaryotic cell.

Energy is captured and extracted by cells' chloroplasts and mitochondria. (pages 92-93)
- Chloroplasts and mitochondria contain their own DNA and may have been independent bacterial cells at one time.

The cytoskeleton is the framework that supports the cell. (pages 94-96)
- Intermediate filaments, microtubules and microfilaments are components of the cytoskeleton.

Bacteria, plants, fungi, and many protists have cell walls that protect and support their cells. (page 97)

All cells transport water and other molecules across their plasma membranes. (pages 98-101)
- Molecules move randomly through the cell by diffusion. Molecules diffuse down concentration gradients, which equalizes concentrations.
- Osmosis occurs when water molecules are able to move through a membrane and other polar molecules are blocked.
- Endocytosis and exocytosis are the bulk passage of food and liquids into and out of a cell.

Selective transport of molecules enables cells to control what enters and leaves a cell. (pages 102-105)
- Facilitated diffusion uses a special carrier protein to move molecules from high to low concentrations.
- Active transport uses other channels to move molecules to an area of greater concentration and requires the expenditure of ATP.

Key Terms Matching

1. _____ Cell	a.	The movement of molecules due to kinetic energy. No membrane involved.
2. _____ Surface-to-volume ratio	b.	Made of intermediate filaments microtubules, and micro filaments.
3. _____ Cytoplasm	c.	All organisms except bacteria are made of this type of cell.
4. _____ Plasma membrane	d.	A phospholipid bilayer with embedded proteins surrounding a cell's cytoplasm.
5. _____ Lipid bilayer	e.	The site of cellular respiration.
6. _____ Transmembrane protein	f.	Results in the movement of water across a semipermeable membrane.
7. _____ Prokaryotes	g.	A general term for the movement of substances out of the cell.
8. _____ Eukaryotes	h.	This influences the maximum size of cells.
9. _____ Organelles	i.	Made of rRNA and protein – 2 subunits.
10. _____ Cytoskeleton	j.	Life's most basic unit of structure and function.
11. _____ Nucleus	k.	Forms when phospholipids are placed in water.
12. _____ Ribosome	l.	Simple, unicellular organisms with no nucleus or organelles.
13. _____ Mitochondria	m.	A general term for the movement of substances into the cell.

14. _____ Diffusion   n. These span the entire cell membrane in eukaryotic cells.
15. _____ Osmosis   o. Chromosomes are located here.
16. _____ Endocytosis   p. Semifluid matrix in which organelles are found.
17. _____ Exocytosis   q. Membrane-bound compartments with specialized functions.

5.1   Cells (Pages 78-81)

18. As cells increase in size, the _____ increases at a faster rate than does the _____ .
19. Cells arise only from _____.

5.2   The Plasma Membrane (Pages 82-84)

20. Persons having the disease called _____ experience excessive mucus secretion that clogs the lungs.
21. Phospholipid molecules have two regions: a polar _____ region and a nonpolar _____ region.
22. _____ molecules bound to the surface of the cell membrane serve as identification markers.
23. One role of surface proteins is to _____.

5.3   Prokaryotic Cells (Page 85)

24. Bacterial DNA is located in
    a. the nucleus.   b. the nucleoid region.   c. the nucleolus.   d. the nucleoli.   e. both a and b.

25. Which of the following is found in BOTH prokaryotic and eukaryotic cells?
    a. Mitochondria   b. A nucleus   c. Chloroplasts   d. Ribosomes   e. Endoplasmic reticulum

5.4   Eukaryotic Cells (Pages 86-87)

26. Label the parts of this animal cell (a – e).

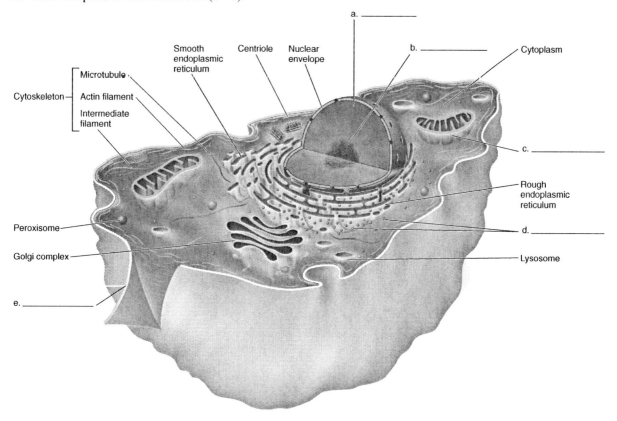

5.5   The Nucleus: The Cell's Control Center (Pages 88-89)

True (T) or False (F) Questions
If you believe the statement to be false, rewrite the statement as a true one.

27. The nuclear envelope is a bilayered structure which has numerous pores.
    Answer: _____ Restatement: _____
    _____

28. The gelatinous fluid inside the nucleus is called nucleoplasm.
    Answer: _____ Restatement: _____
    _____

29. Prior to replication the DNA in the nucleus is in the form of a tangled mass called chromatin.
    Answer: _____ Restatement: _____
    _____

5.6   The Endomembrane System (Pages 90-91)

30. _____ Ribosomes
31. _____ Cell wall
32. _____ Organelle
33. _____ Lysosome
34. _____ Golgi complex
35. _____ Endoplasmic reticulum
36. _____ Mitochondrion
37. _____ Chloroplast
38. _____ Chromatin
39. _____ Microtubule
40. _____ Peroxisome
41. _____ Vesicle
42. _____ Central vacuole
43. _____ Rough ER
44. _____ Smooth ER

a. Have their own DNA and are the site of oxidative metabolism.
b. The site of protein synthesis; made of RNA and protein.
c. This structure surrounds the plasma membrane in plant cells.
d. A network of internal membranes.
e. Form the cytoskeleton.
f. Function in the digestion of cell debris and cell death.
g. Plant organelle, contains photosynthetic pigments.
h. Stacks of flattened vesicles.
i. Membrane-bound structure with a specialized function.
j. Long, threadlike molecule of DNA.
k. Storage center for water and other materials in a plant cell.
l. Very few ribosomes are attached to this structure.
m. Production of proteins and storage of cellular products.
n. A small, membrane-bound sac.
o. Detoxification of potentially harmful molecules.

5.7   Organelles That Contain DNA (Pages 92-93)

45. Label the parts of the mitochondrion (a – d).

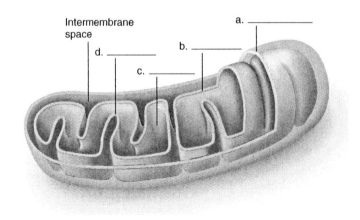

46. Label the parts of the chloroplast (a – e).

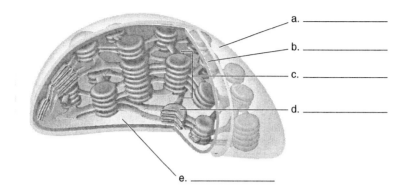

a. _____

b. _____

c. _____

d. _____

e. _____

5.8   The Cytoskeleton: Interior Framework of the Cell (Pages 94-96)

47. Although animal cells lack a cell wall, their structure is maintained by the _____.

48. Microfilaments are made of the protein _____, while microtubules are made of the protein _____ .

49. Cellular crawling is observed in the _____ response and is accomplished by _____ cells.
    a. flight, nerve
    b. healing, red blood
    c. breeding, sex
    d. inflammation, white blood
    e. feeding, amoeboid

50. The movement of chromosomes during mitosis is due to the
    a. shortening of microtubules to which they are attached.
    b. lengthening of microtubules to which they are attached.
    c. shortening of microfilaments to which they are attached.
    d. lengthening of microfilaments to which they are attached.
    e. attraction of DNA to the centrioles.

51. 9 + 2 arrangement refers to the
    a. structure of prokaryotic flagella.
    b. structure of eukaryotic flagella.
    c. structure of prokaryotic cilia.
    d. structure of eukaryotic cilia.
    e. both b and d.

5.9   Outside the Plasma Membrane (Page 97)

52. Plant cell walls are composed of
    a. chitin.   b. cellulose.   c. lipids.   d. all of the above.

53. The protein _____ is abundant in the extracellular matrix.

5.10   Diffusion and Osmosis (Pages 98–99)
True (T) or False (F) Questions
If you believe the statement to be false, rewrite the statement as a true one.

54. Blood cells placed in pure water will burst because there is a net movement of water out of the cell.
    Answer: _____ Restatement: _____
    _____

55. Diffusion is the movement of water molecules across a semipermeable membrane which continues until equilibrium has been reached.
    Answer: _____ Restatement: _____
    _____

5.11   Bulk Passage Into and Out of the Cell (Pages 100-101)

56. Human egg cells obtain nutrients while maturing when surrounding cells
    a.   are ingested by the egg cell.
    b.   secrete nutrients which the egg cell takes in by pinocytosis.
    c.   lyse to release their contents.
    d.   encase them in a coating of nutrients.
    e.   deliver carbohydrates and proteins.

57. _____ is a term that means "cell eating."

58. LDL is taken into the cell by
    a. diffusion.   b. osmosis.   c. receptor-mediated endocytosis.   d. none of the above

5.12   Selective Permeability (Pages 102-105)

59. The difference between diffusion and facilitated diffusion is that_____.

60. _____ is the movement of molecules across a membrane and against a concentration gradient. _____ is the energy used to fuel this process.

61. More than one-third of the energy expended by a human cell is used to drive the _____.

62. Cells obtain information about their surroundings from a battery of _____ that protrude from the cell membrane.

Chapter Test

1. Which of the following would be a problem for a cell if it were able to grow to twice its normal size?
   a. water retention   d. intracellular communication
   b. salt balance   e. communication between different types of cells
   c. reproduction

2. The cell theory states that
   a. cells arise spontaneously.   d. organelles can survive when separated from cells.
   b. cells arise from preexisting cells.   e. both b and c.
   c. cells are the smallest unit of life.

3. Which portion of a phospholipid molecule would be facing the environment outside of the cell?
   a. the polar head   d. the nonpolar tail
   b. the nonpolar head   e. both a and d
   c. the polar tail

4. Which portion of a phospholipid molecule would be facing the interior of the cell?
   a. the polar head
   b. the nonpolar head
   c. the polar tail
   d. the nonpolar tail
   e. both b and d

5. Which portion of a phospholipid molecule is repelled by water?
   a. the polar head
   b. the nonpolar head
   c. the polar tail
   d. the nonpolar tail
   e. none of the above

6. Membrane proteins
   a. exist in fixed positions within the plasma membrane.
   b. float within the plasma membrane.
   c. are completely polar.
   d. are always large enough to span the entire width of the plasma membrane.
   e. are absent in animal cells.

7. How might viruses play a role in curing cystic fibrosis?
   a. Viruses could remove the defective gene from lung cells in persons with cystic fibrosis.
   b. Viruses could repair the defective gene in lung cells in persons with cystic fibrosis.
   c. Viruses could identify the defective gene for later removal.
   d. Viruses could deliver a copy of the normal gene to lung cells in persons with cystic fibrosis.
   e. Viruses cannot possibly play a role in curing cystic fibrosis.

8. Which of the following exhibit the 9 + 2 structure of microtubules?
   a. flagella
   b. cilia
   c. sensory hairs in the human ear
   d. basal bodies
   e. all of the above

9. Centrioles may have originated as
   a. a unique arrangement of microtubules found in some prokaryotic species.
   b. protein fibers produced by ribosomes in some eukaryotic cells.
   c. protozoans that formed an endosymbiotic relationship with primitive aerobic prokaryotes.
   d. spirochetes that became involved in an endosymbiotic relationship.
   e. locomotor structures.

10. Access to the nucleus is regulated by
    a. specialized proteins within the nuclear pore.
    b. the rate of osmosis.
    c. the amount of ATP available for active transport.
    d. hormones.
    e. the nucleolus.

11. A busy student pours herself a cup of coffee and adds a sugar cube. A moment later the phone rings so she doesn't have time to stir it. Ten minutes later she returns to her coffee, and when she tastes it, it is obvious that the sugar has dissolved and that the coffee is uniformly sweet. What happened?
    a. Osmosis has caused the sugar molecules to move down the concentration gradient until they were evenly distributed in the coffee.
    b. Facilitated diffusion has occurred and has caused the sugar molecules to dissolve in the coffee.
    c. Diffusion has occurred in the coffee, causing the sugar to become evenly distributed.
    d. Active transport has occurred using the heat in the coffee as a source of energy.
    e. None of the above.

12. The sugar molecules in the coffee mentioned in question 11 bounce and jostle each other because they have
    a. kinetic energy.   b. potential energy.   c. static energy.   d. random energy.   e. occasional energy.

13. Water molecules crossing a plasma membrane do so by
    a. osmosis.
    b. diffusion.
    c. facilitated diffusion.
    d. active transport.
    e. transport by the sodium-potassium pump.

14. Which of the following is a requirement for osmosis to occur?
    a. a semipermeable membrane
    b. the sodium-potassium pump
    c. a specific receptor protein
    d. an energy source
    e. all of the above

15. What advantage would a plant cell have over an animal cell if both were placed in pure water?
    a. The plant cell would resist shrinking in pure water.
    b. The plant cell would resist bursting in pure water.
    c. The plant cell would reproduce at a faster rate in pure water.
    d. Both a and c.
    e. None of the above.

16. The release of insulin from certain human cells is an example of
    a. endocytosis.   b. exocytosis.   c. phagocytosis.   d. pinocytosis.   e. both a and b.

17. An amoeba is engulfing bacterial cells by surrounding them with a pseudopod. This is an example of
    a. endocytosis.   b. exocytosis.   c. phagocytosis.   d. pinocytosis.   e. both a and c.

18. The removal of waste products from the amoeba is an example of
    a. endocytosis.   b. exocytosis.   c. phagocytosis.   d. pinocytosis.   e. both a and c.

19. Diffusion of molecules across biological membranes that requires channels is called
    a. active transport.   b. the sodium-potassium pump.   c. facilitated diffusion.
    d. osmosis.   e. none of the above.

20. The sodium-potassium pump functions to
    a. remove sodium and potassium ions from the cell.
    b. pump sodium and potassium ions into the cell.
    c. pump potassium ions into the cell and sodium ions out of the cell.
    d. pump sodium ions into the cell and potassium ions out of the cell.
    e. pump sodium and potassium ions into the cell and calcium out of the cell.

21. A liver cell would detect the presence of insulin by
    a. means of receptor proteins on its surface.
    b. means of receptor carbohydrates on its surface.
    c. producing insulin-detection binding molecules.
    d. the reduction of glucose in the external environment.
    e. both a and b.

Additional Study Help

Visit the ARIS (Assessment, Review, and Instruction System) site at aris.mhhe.com for quizzes, animations, and other study tools.

6 Energy and Life

Key Concepts Outline

Energy is the ability to do work; energy being used is kinetic energy, while energy being stored for future use is potential energy. (page 110)
- In cells, chemical reactions occur to make or break chemical bonds, which stores or releases energy, respectively.

The amount of energy in the universe remains constant, but energy can be transformed from one form to another. When energy is transformed some is lost as heat. (page 111)
- Heat energy is unable to do the work of cells, so new energy from the sun constantly entering the system replaces the energy lost as heat.

All chemical reactions require an input of energy, the activation energy, to start the reaction. (page 112)
- Catalysts lower the energy of activation, which can make a reaction proceed more quickly.

Enzymes are biological catalysts used by cells to initiate particular chemical reactions. (pages 113-114)
- An enzyme is specific for a particular reactant or substrate; the reactant fits the active site of an enzyme.
- Enzymes are not affected by the reactions they facilitate and are available to be used again.
- Temperature and pH affect the activity of enzymes.

Cells can control enzyme function by altering the shape of an enzyme. (page 115)
- Signal molecules can bind to allosteric enzymes to control the enzyme's activity.
- Feedback inhibition occurs when the product of a reaction inhibits the enzyme's activity.

Energy from the sun or food is converted by cells into ATP. (pages 116-118)
- ATP is broken down inside cells, enabling them to perform their functions.

Key Terms Matching

1. _____ Chemical reaction
2. _____ Activation energy
3. _____ Reactants
4. _____ Enzyme
5. _____ ATP

a. The energy required to cause a reaction to go to completion.
b. Has a third bond that is especially energy rich.
c. Reaction-starting molecules.
d. Proteins that catalyze chemical reactions.
e. The making or breaking of chemical bonds.

6.1 The Flow of Energy in Living Things (Page 110)

6. The capacity to do work is _____, the study of which is _____.

7. The unit of heat used most often in biology is the _____.

6.2 The Laws of Thermodynamics (Page 111)

True (T) or False (F) Questions
If you believe the statement to be false, rewrite the statement as a true one.

8. The quantity of energy is constantly increasing as the universe continues to expand.
   Answer: _____ Restatement: _____
   _____

9. The Second Law of Thermodynamics concerns the amount of energy in the universe.
   Answer: _____ Restatement: _____
   _____

10. The universe is constantly becoming more disordered.
Answer: _____ Restatement: _____

_____

6.3   Chemical Reactions (Page 112)

11. _____ is the lowering of the _____ energy of a chemical reaction.
12. Reactions that release energy are referred to as _____ while those that will not occur unless an outside source of energy is supplied are _____.

6.4   How Enzymes Work (Pages 113-114)

13. The three-dimensional surface on an enzyme to which the reactant binds is called the
    a. substrate-complex site.   b. specific site.   c. active site.   d. reactive site.   e. reactive area.

14. Coenzymes are
    a. metallic side-groups associated with all enzymes.   d. nonprotein reactants.
    b. metallic side-groups associated with nonprotein enzymes.   e. the product of enzyme catalysis.
    c. nonprotein, organic molecules that act as cofactors.

6.5   How Cells Regulate Enzymes (Page 115)

15. Enzymes are sensitive to _____ and _____.
16. Enzymes are regulated by a mechanism called _____.
17. A competitive inhibitor binds to the _____ site.

6.6 ATP: The Energy Currency of the Cell (Pages 116-118)

True (T) or False (F) Questions
If you believe the statement to be false, rewrite the statement as a true one.

18. The three parts of an ATP molecule are a ribose sugar, adenine molecule and three phosphate groups.
    Answer: _____ Restatement: _____

    _____

19. When the terminal phosphate group is broken off of an ATP molecule, a significant quantity of energy is absorbed.
    Answer: _____ Restatement: _____

    _____

20. When an atom loses an electron, it is oxidized
    Answer: _____ Restatement: _____

    _____

Chapter Test

1. Which of the following is an example of a chemical reaction?
   a. breaking the hydrogen bonds between water molecules
   b. forming hydrogen bonds between water molecules
   c. joining amino acids to form a protein molecule
   d. storing fat in cells
   e. all of the above

2. The energy required to initiate a chemical reaction is called
   a. activation energy.   d. primary energy.
   b. initiation energy.   e. initial energy.
   c. catalysis.

26

3. Even when catalysts are available, which of the following is also required for a reaction to occur?
   a. favorable atmospheric conditions
   b. oxygen
   c. energy
   d. zinc
   e. hydrogen

4. In many reactions NAD$^+$ functions to
   a. accept electrons.
   b. transform electrons.
   c. release electrons.
   d. all of the above.
   e. both a and c.

5. A cell might regulate the production of a certain substance by
   a. destroying the enzymes required for a key reaction.
   b. altering the shape of a key enzyme.
   c. becoming dormant when adequate quantities of the product are detected.
   d. destroying excess quantities of the product.
   e. storing enzymes until the product is exhausted.

6. A consequence of removing a phosphate from a molecule of ATP is
   a. the release of some energy.
   b. the consumption of some energy.
   c. the consumption of some oxygen.
   d. the production of cellular waste.
   e. both a and d.

7. The process of lowering the activation energy of a reaction is called
   a. endergonic.
   b. exergonic.
   c. catalysis.
   d. none of the above.

8. The site on the enzyme surface where the reactant fits is called the
   a. allosteric site.
   b. active site.
   c. binding site.
   d. inhibitory site.

9. Most human enzymes work best within a pH range of
   a. 1–3.
   b. 3–6.
   c. 6–8.
   d. 8–10.

10. Noncompetitive inhibitors bind to
    a. the active site.
    b. the allosteric site.
    c. the substrate.
    d. none of the above.

11. Most energy exchanges in cells involve the conversion of
    a. ATP to ADP.
    b. ATP to AMP.
    c. ADP to AMP.
    d. ADP to ATP

Additional Study Help

Visit the ARIS (Assessment, Review, and Instruction System) site at aris.mhhe.com for quizzes, animations, and other study tools.

7 Photosynthesis: Acquiring Energy From the Sun

Key Concepts Outline

Energy ultimately reaches organisms via photosynthesis. (pages 122-123)
- Energy from sunlight is captured and used to make ATP and NADPH.
- ATP and NADPH are used to convert CO_2 in the air into organic molecules.
- The chloroplast is the site of the three stages of photosynthesis.

Packets of light energy called photons are absorbed by pigments. (pages 124-125)
- Chlorophyll is the main pigment in plants that absorbs light energy; it reflects green wavelengths of light, making plants appear green.

Light-dependent reactions occur in the membranes of the thylakoids. (pages 126-127)
- The five stages of the light-dependent reactions are: capturing light, exciting an electron, electron transport, making ATP, and making NADPH.

Noncyclic photophosphorylation produces ATP and NADPH with two photosystems. (pages 128-129)
- Photosystem II generates high-energy electrons used to synthesize ATP; those electrons are then passed to photosystem I where they're used to make NADPH.

The Calvin cycle occurs in the chloroplasts' stroma and follows the light-dependent reactions and produces organic molecules. (pages 130-131)
- ATP and NADPH produced during the light-dependent reactions are used to create carbohydrates; ADP and NADP+ are recycled.

The light-independent reactions performed by plants adapted to hot weather vary from the Calvin cycle. (pages 132-134)
- Some plants adapted to higher temperature perform C_4 photosynthesis; CAM plants' stomata open during the night and they fix CO_2 at night.

Key Terms Matching

1. _____ Photosynthesis	a. Molecules that absorb light.
2. _____ Light-independent reaction	b. Location of light-independent reactions.
3. _____ Thylakoid	c. A three-carbon molecule is the first product of these reactions.
4. _____ Pigments	d. Process of capturing sunlight to make energy.
5. _____ Carotenoids	e. Location of light-dependent reactions.
6. _____ Chlorophyll	f. Stage of photosynthesis that does not require light directly.
7. _____ Stroma	g. Primary pigment of photosynthesis.
8. _____ Calvin cycle	h. Capture blue and green light and reflect orange and yellow light.

7.1   An Overview of Photosynthesis (Pages 122-123)

9. Photosynthesis is carried out in the _____ of plants.
    a. mitochondria
    b. plasma membrane
    c. chloroplast
    d. nucleus

10. Chlorophyll pigments are grouped together in a network called _____.

7.2   How Plants Capture Energy from Sunlight (Pages 124-125)

11. Light consists of tiny packets of energy called _____.

7.3   Organizing Pigments into Photosystems (Pages 126-127)

Refer to Figure 7.7 to match the reaction center to the corresponding photosystem.

12. _____ P_{700} a. photosystem I
13. _____ P_{680} b. photosystem II

7.4   How Photosystems Convert Light to Chemical Energy (Pages 128-129)

14. The process of producing ATP during photosynthesis is called _____.
15. The light-dependent reaction produces _____ and _____ that are later used in the Calvin cycle.
16. The light-dependent reaction produces _____ as a waste product.

7.5   Building New Molecules (Pages 130-131)

17. _____ Calvin cycle   a. Where C_3 photosynthesis takes place.
18. _____ Stroma   b. The process of "fixing carbon."

7.6   Photorespiration: Putting the Brakes on Photosynthesis (Pages 132-134)

19. _____ Photorespiration   a. An adaptation seen in plants living in hot climates.
20. _____ C_4 photosynthesis   b. This occurs when plants must close their stomata.

Chapter Test

1. In plants photosynthesis occurs in the
    a. central vacuole of most cells.
    b. cell wall.
    c. leaves only.
    d. chloroplasts.
    e. vascular tissue.

2. The actual formation of organic molecules from atmospheric carbon dioxide requires
    a. light.   b. dark.   c. ATP.   d. NAD^+.   e. NADH.

3. In a high school biology class, students observe that plants grown with a green filter in front of the light source grow better than plants with a red filter in front of the light source. Why is this?
    a. Plants do not absorb the red wavelengths of light for photosynthesis.
    b. Plants do not absorb the green wavelengths of light for photosynthesis.
    c. Plants reflect the red wavelengths of light during photosynthesis.
    d. Plants cannot reflect light during photosynthesis.
    e. both a and b.

4. The presence of carotenoid pigments can be demonstrated by
    a. observing mature leaves during the hot summer months.
    b. bleaching leaves to remove all of the chlorophyll.
    c. observing leaves in the cool fall months.
    d. using a light microscope.
    e. both a and b.

5. Chlorophyll *a* and chlorophyll *b* absorb slightly different wavelengths of light primarily due to differences in
   a. the primary chlorophyll molecule.
   b. "side groups" attached to the molecule.
   c. light quality.
   d. the carotenoid pigments available.
   e. none of the above.

6. Oxygen is produced during photosynthesis when
   a. the oxygen is removed from carbon dioxide to make carbohydrate.
   b. hydrogen is added to carbon dioxide to make carbohydrate.
   c. water molecules are split to provide electrons for photosystem I.
   d. water molecules are split to provide electrons for photosystem II.
   e. oxygen is added to chlorophyll.

7. Plants living in the desert conserve water at least in part by
   a. closing stomata when the temperature exceeds a critical point.
   b. increasing their uptake of water during the summer months.
   c. growing in groups with other related plants.
   d. absorbing increased amounts of oxygen in the summer.
   e. both a and c.

8. A consequence of increased oxygen concentration in the leaves of a plant is
   a. decreased absorption of light by chlorophyll.
   b. reduction in RuBP concentration in leaves.
   c. growing in groups with other related plants.
   d. more efficient photosynthesis.
   e. overproduction of carotenoid pigments.

9. _____ is split during the light-dependent reaction so a continuous flow of electrons through the photosystems is maintained.
   a. CO_2 d. O_2
   b. H_2O e. $C_6H_{12}O_6$
   c. ATP

10. _____, the products of the light-dependent reaction, are used to synthesize glucose during the light-independent reaction
   a. H_2O and CO_2 d. O_2 and ATP
   b. ATP and NADPH   e. O_2 and CO_2
   c. RuBP and OAA

Additional Study Help

Visit the ARIS (Assessment, Review, and Instruction System) site at aris.mhhe.com for quizzes, animations, and other study tools.

8   How Cells Harvest Energy From Food

Key Concepts Outline

Energy is obtained by oxidizing foodstuffs via a series of events called cellular respiration. (pages 138-139)
- The first stage of cellular respiration is glycolysis, during which coupled reactions produce ATP in the cytoplasm; oxygen is not required.
- Stage two involves a cycle of chemical reactions called the Krebs cycle and the use of an electron transport chain to produce a great deal of ATP; this takes place in the mitochondria and requires oxygen.
- Ultimately, oxidative cellular respiration consumes O_2 and releases CO_2.

Glycolysis is performed by all living organisms and is followed by fermentation or some other anaerobic respiration in the absence of oxygen. (pages 140-143)
- There is a net gain of 2 molecules of ATP, 2 NADH molecules and 2 molecules of pyruvate from glycolysis.
- Two kinds of fermentation produce lactate or ethanol and the NAD^+ needed to continue glycolysis.

Acetyl CoA production and the Krebs cycle follow glycolysis in the presence of oxygen. (pages 144-147)
- The oxidation of pyruvate in the mitochondria produces acetyl CoA, CO2 and NADH.
- Acetyl CoA enters the Krebs cycle, which produces CO_2, the electron carriers NADH and $FADH_2$ and ATP.

An electron transport chain in the mitochondria accepts electrons and produces numerous ATP molecules. (pages 148-149)
- Electrons are passed from NADH and $FADH_2$ to the membrane associated molecules, and ATP molecules are synthesized via chemiosmosis.
- Oxygen combines with protons and electrons to form water.

Proteins and fats in your food are also important sources of energy. (page 150)
- Amino acids from proteins and fatty acids from fats can be converted to products that can be oxidized during cellular respiration.

Key Terms Matching

1. _____ Cellular respiration	a. In yeast pyruvate is converted to ethanol.
2. _____ Glycolysis	b. First stage of cellular respiration.
3. _____ Substrate-level phosphorylation	c. Membrane molecules that accept electrons.
4. _____ Mitochondria	d. Mitochondrial reactions that follow glycolysis if O_2 is present
5. _____ Krebs cycle	e. ATP formed by diffusion force of protons.
6. _____ Fermentation	f. Oxygen-requiring events yielding numerous ATP molecules
7. _____ Chemiosmosis	g. Location of Krebs cycle and electron transport chain
8. _____ Electron transport chain	h. Transfer of a high-energy phosphate group to ADP

8.1   Where Is the Energy in Food? (Pages 138-139)

9. In aerobic respiration _____ serves as the final electron acceptor while _____ is the final electron acceptor in anaerobic respiration.

10. Anaerobic respiration is performed by some primitive bacteria such as the _____ and the _____.

8.2   Using Coupled Reactions to Make ATP (Pages 140-142)
True (T) or False (F) Questions
If you believe the statement to be false, rewrite the statement as a true one.

11. The step that occurs before respiration takes place is called fermentation.
     Answer: _____ Restatement: _____

_____

8.3 Harvesting Electrons from Chemical Bonds (Pages 144-147)

12. _____ is the end product of glycolysis. It is oxidized into _____, which then enters the Krebs cycle.
    a. NADH, pyruvate
    b. Pyruvate, NADH
    c. Acetyl CoA, pyruvate
    d. Pyruvate, acetyl CoA
    e. CO_2, NADPH

8.4 Using Electrons to Make ATP (Pages 148-149)

13. Which of the following is the best description of cytochrome c?
    a. Cytochrome c is another name for ubiquinone, a protein carrier molecule which functions in the electron transport chain.
    b. Cytochrome c is the source of energy for the proton pumps found in mitochondria.
    c. Cytochrome c is a type of photosynthetic molecule that is unique to several genera of prokaryotes.
    d. Cytochrome c is a carrier molecule in the electron transport chain.
    e. Cytochrome c functions by binding with the coenzyme NADP to initiate aerobic respiration.

14. In mitochondria the Krebs cycle takes place in
    a. the outermost membrane.
    b. the matrix.
    c. pores located throughout the inner membrane.
    d. all parts of the organelle.
    e. The Krebs cycle does not occur in mitochondria.

8.5 Glucose Is Not the Only Food Molecule (Page 150)

15. _____ from proteins can be converted into molecules that take part in the Krebs cycle.
    a. Fatty acids   b. Nucleotides   c. Amino acids   d. Phosphates   e. Monosaccharides

16. Fatty acids from _____ can be converted to acetyl groups that combine with coenzyme A to form acetyl CoA.
    a. polysaccharides   b. proteins   c. ATP   d. lipids   e. nucleic acids

Chapter Test

1. CO_2 is produced during
   a. lactate fermentation.   d. acetyl CoA production and the Krebs cycle.
   b. glycolysis.   e. all of these.
   c. chemiosmosis.

2. Fermentation is beneficial to cells because it allows them to regenerate the _____ needed for glycolysis to continue.
   a. $FADH_2$ b. water   c. NAD^+ d. coenzyme A   e. pyruvate

3. The final electron acceptor in the electron transport system is
   a. oxygen.   d. glucose.
   b. carbon dioxide.   e. lactic acid (lactate).
   c. NADH.

4. Per molecule of glucose,
   a. 2 pyruvates are produced.   d. 6 molecules of O_2 are produced.
   b. 36 molecules of CO_2 are produced.   e. all of the above.
   c. there is 1 turn of the Krebs cycle.

5. Fermentation occurs in the
   a. mitochondrion.
   b. cytoplasm.
   c. ribosomes.
   d. chloroplast.
   e. nucleus.

6. In the absence of glucose
   a. fermentation occurs.
   b. lactate/lactic acid is produced.
   c. fatty acids, glycerol or amino acids are used to make ATP.
   d. ATP synthesis halts.
   e. only 2 ATP are produced.

7. Acetyl CoA formation and the Krebs cycle take place in/at the
   a. nucleus.
   b. ribosomes.
   c. mitochondria.
   d. cytoplasm.
   e. chloroplast.

8. Which of these products of the preparatory steps and Krebs cycle are most likely to end up in a plant's glucose?
   a. NADH
   b. CO_2
   c. coenzyme A
   d. acetyl CoA
   e. H_2O

9. Which of the following reactions is most likely to occur in the cytoplasm of a eukaryotic cell and in the absence of oxygen?
   a. chemiosmosis
   b. acetyl CoA formation
   c. Krebs cycle
   d. glycolysis
   e. photosynthesis

10. The overall reaction for aerobic respiration is summarized as
    a. $C_6H_{12}O_6 + 6O_2 + 6H_2O \rightarrow 6CO_2 + 12H_2O + energy$
    b. $C_6H_{12}O_6 + 6H_2O \rightarrow 6CO_2 + 12O_2 + energy$
    c. $C_6H_{12}O_6 + 6CO_2 + ATP \rightarrow 6O_2 + 12H_2O + energy$
    d. none of the above

Additional Study Help

Visit the ARIS (Assessment, Review, and Instruction System) site at aris.mhhe.com for quizzes, animations, and other study tools.

9 Mitosis

Key Concepts Outline

Prokaryotic cells divide by binary fission. (page 154)
- After the DNA has been copied, the cell elongates and forms two daughter cells by splitting into two equal halves.

Eukaryotic cells contain more DNA and have a more complex method of dividing than prokaryotic cells. (page 155)
- Interphase has three phases: cell growth occurs in G_1, DNA replicates and produces two copies of each chromosome in S, and preparation for division occurs in G_2.
- During mitosis the chromosome copies are separated so the two daughter cells created during cytokinesis will have a copy of each chromosome.

Chromosomes are composed of DNA and proteins and exist in somatic cells in pairs called homologous chromosomes. (pages 156-157)
- Diploid cells have two of each kind of chromosome; human somatic cells have 46 chromosomes.
- The two copies of each homologous chromosome are called sister chromatids that are joined by a centromere.

Eukaryotic cell division begins with interphase, continues with four stages of mitosis and ends with cytokinesis. (pages 158-160)
- The cell prepares for cell division during interphase.
- Mitosis follows interphase and has four stages: prophase, metaphase, anaphase, and telophase.
- The cytoplasm is divided during cytokinesis, resulting in two daughter cells.

The cell cycle of eukaryotic cells is controlled at three checkpoints. (page 151)
- There are checkpoints during the G_1 and G_2 phases of interphase and at metaphase during mitosis.

Cancer is a disease in which the regulatory controls that normally restrain cell division are disrupted. (page 162)
- A variety of environmental factors, including ionizing radiation, chemical mutagens, and viruses, have been implicated in causing cancer.

Many cancers are associated with mutations that disable key elements of the G_1 checkpoint. (page 163)
- The gene *p*53 plays an important role in checking for damaged DNA during G_1.

Key Terms Matching

1. _____	Binary fission	a. The formation of new tumors.
2. _____	Chromosome	b. An arrangement of paired homologous chromosomes.
3. _____	Mitosis	c. Includes interphase as well as the stages of mitosis.
4. _____	Cell cycle	d. Nonreproductive cells.
5. _____	Cancer	e. Having two sets of homologous chromosomes.
6. _____	Metastases	f. The process by which prokaryotic cells divide.
7. _____	Diploid	g. Uncontrolled cellular division.
8. _____	Somatic cells	h. Linkage site that joins sister chromatids.
9. _____	Centromere	i. DNA and associated proteins.
10. _____	Karyotype	j. A diploid eukaryotic cell divides by this process.

9.1 Prokaryotes Have a Simple Cell Cycle (Page 154)
True (T) or False (F) Questions
If you believe the statement to be false, rewrite the statement as a true one.

11. The term "simple cell cycle" is a reference to the production of two identical diploid eukaryotic cells.
    Answer: _____ Restatement: _____

12. Binary fission begins after DNA replication is completed.
    Answer: _____ Restatement: _____

9.2 Eukaryotes Have a Complex Cell Cycle (Page 155)

13. During interphase,
    a. chromosomes are lined up on the equator.
    b. chromosomes have separated into sister chromatids.
    c. the nuclear envelope breaks down.
    d. cytoplasm divides.
    e. DNA replicates.

14. During the meta phase of mitosis,
    a. DNA is replicated.
    b. homologous chromosomes line up on the equator.
    c. DNA is uncoiled and is in use.
    d. spindle fibers are bound to chromosomes.
    e. preparation for cell division begins.

9.3 Chromosomes (Pages 156–157)

15. A(n) _____ is a photograph of an individual's chromosomes that are cut out and arranged in _____.

16. Chromosomes were first discovered by _____ in 1882.

17. Cells having two of each type of chromosome are called _____ cells.

9.4 Cell Division (Pages 158-160)

True (T) or False (F) Questions
If you believe the statement to be false, rewrite the statement as a true one.

18. The spindle is a network of protein cables extending from centrioles at each pole of the cell during mitosis.
    Answer: _____ Restatement: _____
    _____

19. Two diploid daughter cells are the result of mitosis.
    Answer: _____ Restatement: _____
    _____

9.5 Controlling the Cell Cycle (Page 161)

20. The G_1 checkpoint is the time when
    a. sister chromatids separate and move to opposite poles of the cell.
    b. the decision is made to divide, delay or enter a resting stage.
    c. cytokinesis begins.
    d. cancerous cells begin unchecked division.
    e. none of the above.

21. The cell cycle of eukaryotes is accessed and controlled by feedback
    a. at three checkpoints.
    b. provided by specialized cells of the immune system.
    c. at the G_1 checkpoint. If all is well, mitosis continues unchecked to completion.
    d. during the S phase.
    e. both b and c.

9.6 What Is Cancer? (Page 162)

22. Cancerous cells that leave the original tumor and form tumors at other sites are called _____.

23. Two agents that can cause cells to become cancerous are _____ and _____.

9.7   Cancer and Control of the Cell Cycle (Page 163)
True (T) or False (F) Questions
If you believe the statement to be false, rewrite the statement as a true one.

24. p53 is the strain of virus most often responsible for causing cells to become cancerous.
    Answer: _____ Restatement: _____
    _____

25. A disabled G$_1$ checkpoint is associated with many cancers.
    Answer: _____ Restatement: _____
    _____

A Closer Look: Curing Cancer (Pages 164-165)

26. Tumors release substances into the bloodstream that result in _____ and aid in providing additional nutrients.
     a. increased cell division
     b. death of surrounding cells
     c. angiogenesis
     d. decreasing local blood supply
     e. both a and d are true

27. Epidermal growth factor receptors
     a. are now produced as a molecular therapy for the treatment of liver cancer.
     b. are used by cells to remove the "brake" the cell uses to restrain cell division.
     c. are overproduced in over 20% of breast cancers.
     d. amplify the signal in cytoplasm telling the cell to divide.
     e. regulate the three checkpoints that control cell division.

Chapter Test

1. The structure of eukaryotic chromosomes is in part maintained by
    a. covalent bonding between the strands of the double helix.
    b. ionic bonding between the strands of the double helix.
    c. winding the DNA molecule around a core of histone protein.
    d. the centromere.
    e. microtubules present in the cytoplasm.

2. Sister chromatids are
    a. pairs of heterozygous chromosomes.
    b. only seen in prokaryotic cells.
    c. formed only during meiosis.
    d. duplicated DNA held together by a centromere.
    e. both a and d.

3. DNA binds with histone proteins because
    a. the combination of the two creates a positively charged structure.
    b. the negatively charged histone is attracted to the negatively charged DNA.
    c. the combination of the two creates a complex with no net charge.
    d. that is the only way that DNA replication can be stimulated.
    e. none of the above.

4.   Human skin cells are
     a. dead when mature.   b. haploid.   c. diploid.   d. capable of becoming germ cells.
     e. produced by meiosis.

5. While viewing stained onion root tip cells you notice that distinct chromosomes cannot be observed in most of the cells. This is because most of the cells
   a. were destroyed while preparing the slide.
   b. must be in interphase.
   c. are in metaphase.
   d. must be undergoing meiosis.
   e. are in stasis.

6. The S phase of mitosis involves synthesis of
   a. cellular products for export from the cell.
   b. cellular membranes and organelles.
   c. DNA.
   d. the cell wall in plant cells.
   e. both a and d.

7. The majority of the cell cycle is taken up by
   a. interphase.
   b. prophase.
   c. metaphase.
   d. anaphase.
   e. telophase.

8. The nuclear envelope reforms in the daughter cells during
   a. interphase.
   b. prophase.
   c. metaphase.
   d. anaphase.
   e. telophase.

9. The shortest stage of mitosis is
   a. interphase.
   b. prophase.
   c. metaphase.
   d. anaphase.
   e. telophase.

10. Mutations _____ are associated with many cancers.
    a. preventing the initiation of metaphase
    b. disabling key elements of the G_1 checkpoint
    c. of the cell membrane
    d. caused by viral infections
    e. both a and c

Additional Study Help

Visit the ARIS (Assessment, Review, and Instruction System) site at aris.mhhe.com for quizzes, animations, and other study tools.

10   Meiosis

Key Concepts Outline

A cycle of reproduction involves meiosis and fertilization. (page 170)
- Haploid cells fuse during sexual reproduction and form a diploid zygote.

All sexually reproducing organisms have the same basic pattern of alternating between diploid and haploid chromosome numbers. (page 171)
- In animals diploid cells undergo meiosis, which produces haploid gametes.

There are two divisions during meiosis. (pages 172-175)
- Meiosis I begins with interphase, during which DNA is replicated; prophase I, metaphase I, anaphase I and telophase I follow interphase.
- Crossing over between homologous chromosomes occurs during prophase I, and homologous chromosomes are separated during anaphase I.
- Meiosis II has four stages: prophase II, metaphase II, anaphase II and telophase II.

There are two features that differentiate meiosis from mitosis. (pages 176-177)
- The features unique to meiosis are synapsis, the process of forming homologous chromosome complexes and a reduction division during which homologous chromosomes are separated resulting in haploid cells.

Sexual reproduction has a big impact on species' ability to evolve. (page 178)
- Independent assortment, crossing over and random fertilization all contribute to the impact sexual reproduction has on a species evolution.

Key Terms Matching

1. _____ Gametes	a. One set of chromosomes present.
2. _____ Zygote	b. Gametes have different combinations of parental genes.
3. _____ Fertilization	c. Reproduction that doesn't involve the fusion of gametes.
4. _____ Diploid	d. This results in new genetic combinations during meiosis.
5. _____ Haploid	e. Two sets of homologous chromosomes present.
6. _____ Asexual reproduction	f. Cells involved in sexual reproduction.
7. _____ Synapsis	g. Cell formed by the fusion of gametes.
8. _____ Meiosis	h. Cell division that produces four haploid cells.
9. _____ Crossing over	i. Fusion of gametes to form a new cell.
10. _____ Independent assortment	j. Pairing of homologous chromosomes during which crossing over occurs.

10.1   Discovery of Meiosis (Page 170)

11. _____ Gametes	a. Results from fusion of egg and sperm.
12. _____ Syngamy	b. Another name for fertilization.
13. _____ Meiosis	c. Eggs and sperm.
14. _____ Zygote	d. Produces four haploid cells.

10.2   The Sexual Life Cycle (Page 171)

15. The haploid portion of alternation of generations begins when _____ while the diploid portion begins when _____ occurs.

16. All cells other than gametes are called _____ cells.

10.3 The Stages of Meiosis (Pages 172–175)

17. _____ Prophase I   a. Chromosomes gather at the two poles of the cell.
18. _____ Metaphase I   b. Crossing over occurs.
19. _____ Anaphase I   c. Chromosomes align on the central plane.
20. _____ Telophase I   d. One version of each chromosome moves to a pole of the cell.

10.4 Comparing Meiosis and Mitosis (Pages 176-177)

21. Two unique features of meiosis are _____ and _____.

10.5 The Evolutionary Consequences of Sex (Pages 178-179)

22. Diploid human cells have _____ pairs of chromosomes.
23. More than _____ different kinds of gametes can potentially be formed with the number of chromosomes in human cells.

Chapter Test

1. The reduction division associated with meiosis occurs during
   a. interphase.
   b. telophase II.
   c. anaphase I.
   d. prophase I.
   e. metaphase II.

2. Sister chromatids separate from each other during this/these phases.
   a. anaphase of mitosis
   b. anaphase I
   c. anaphase II
   d. a and c are both correct.
   e. a, b and c are all correct.

3. After meiosis I, you will observe
   a. 2 cells, both diploid.
   b. 2 cells, both haploid.
   c. 4 cells, all diploid.
   d. 4 cells, all haploid.

4. Which of the following takes place directly after telophase I?
   a. prophase II   d. DNA replication
   b. metaphase   e. anaphase II
   c. homologous chromosome separation

5. The somatic cells of humans contain _____ chromosomes and the gametes contain _____ chromosomes.
   a. 46, 46   d. 46, 23
   b. 23, 23   e. 46, 92
   c. 23, 46

6. A human zygote contains _____ chromosomes.
   a. 23   b. 46   c. 69   d. 92

7. Meiosis II and mitosis are similar because
   a. homologous chromosomes separate during anaphase.
   b. the resulting cells are all diploid.
   c. each results in four cells produced.
   d. sister chromatids separate during anaphase.
   e. crossing over occurs during prophase.

8. What evolutionary consequences arise from sexual reproduction?
   a. genetic diversity
   b. binary fission
   c. diploid cells
   d. dimorphism
   e. none of the above

9. Crossing over takes place during
   a. anaphase of mitosis.
   b. telophase of mitosis.
   c. meiosis I.
   d. meiosis II.
   e. cytokinesis.

10. Genetic variation in a population is ensured by
    a. crossing over during meiosis.
    b. random fertilization.
    c. independent assortment.
    d. all of the above.
    e. only a and c.

Additional Study Help

Visit the ARIS (Assessment, Review, and Instruction System) site at aris.mhhe.com for quizzes, animations, and other study tools.

11 Foundations of Genetics

Key Concepts Outline

In the 1860s, Gregor Mendel conducted genetic crosses of varieties of pea plants. (pages 184-190)
- Carefully counting the numbers of each kind of offspring, Mendel observed that in crosses of heterozygous parents, one-quarter of the offspring exhibit the recessive trait.
- To explain his results Mendel formulated a theory composed of five hypotheses.

Two laws of genetic inheritance are attributed to Mendel. (page 191)
- The law of segregation says two alleles for a trait separate during gamete production. Half of the gametes will carry one allele and half the other allele.
- The law of independent assortment says the inheritance of genes on different chromosomes are independent of one another.

Genes determine the amino acid sequence of a protein. (pages 192-193)
- The kinds of proteins produced and present in the body determine how the body functions; this is reflected in the phenotype.
- Mutations can change a phenotype by altering a protein's amino acid sequence, which may alter the protein's function.

Some traits don't reflect Mendelian inheritance. (pages 194-199)
- Phenotypes can be controlled by multiple genes. Phenotypes can be affected by alleles that have incomplete dominance or codominance or influence the expression of one another.

Sex linkage research confirmed that genes determining Mendelian traits are located on chromosomes that assort independently when meiosis occurs. (pages 200-201)
- Thomas Hunt Morgan demonstrated that the segregation of the white-eyed trait in Drosophila is associated with the segregation of the X chromosome.

Nondisjunction results when chromosomes do not separate during meiosis, leading to gametes with missing or extra chromosomes. (pages 202-203)
- In humans, the loss of an autosome is invariably fatal; gaining an extra autosome (called trisomy) is also fatal except when it's trisomy 21 or 22.
- Nondisjuction of the sex chromosomes has less serious consequences, although individuals that have them may be sterile.

Some hereditary disorders are relatively common in human populations; others are rare. (pages 204–209)
- Many of the most important disorders are associated with recessive alleles.
- Genetic counseling techniques like amniocentesis are important aids in predicting the likelihood of producing children who express hereditary disorders.

Key Terms Matching

1. _____ Heredity		a. Failure of chromosomes to separate during mitosis.
2. _____ Recessive		b. One gene modifies the expression of another gene.
3. _____ Gene		c. A family tree.
4. _____ Heterozygous		d. All chromosomes in a cell arranged in order of size.
5. _____ Allele		e. An allele having more than one effect on the phenotype.
6. _____ Epistasis		f. Nucleotide sequence specifying a polypeptide.
7. _____ Pleiotropy		g. Disease that results in brain degeneration due to accumulation of lipids.
8. _____ Sex-linked		h. Alternate versions of a gene.
9. _____ Karyotype		i. Symptoms do not usually develop until after 30 years of age.
10. _____ Nondisjunction		j. Traits written in genes.
11. _____ X chromosome		k. Two of them in female humans.
12. _____ Pedigree		l. The two copies of the factor coding for a trait are different.
13. _____ Tay-Sachs disease		m. Trait determined by gene on a sex chromosome.
14. _____ Huntington's disease		n. A trait which is only expressed in a homozygous organism.

11.1 Mendel and the Garden Pea (Pages 184-185)

15. In order to ensure that he had true-breeding plants for his experiments, Mendel
    a. cross-fertilized each variety with another.
    b. let each variety self-fertilize for several generations.
    c. removed the female parts of the flowers to prevent cross-fertilization.
    d. removed the male parts of the flowers to prevent cross-fertilization.
    e. used numerous varieties of garden peas.

16. When Mendel crossed two varieties of true-breeding plants the F_1 generation were
    a. homozygous for one trait or the other.
    b. heterozygous.
    c. of a phenotype midway between the parent plants.
    d. phenotypically similar to the parent plant with the dominant trait.
    e. both b and d.

11.2 What Mendel Observed (Pages 186-187)

17. The particular alleles an organism possesses determine its _____, while its physical appearance is the organism's _____.
18. When Mendel allowed the F_1 generation of plants to self-fertilize the F_2 generation, they showed a ratio of _____.

11.3 Mendel Proposes a Theory (Pages 188-190)

19. An organism having one dominant allele and one recessive allele is said to be _____ for the trait.
20. Albinism and hemophilia are examples of _____ traits.

21. In a particular breed of cats black (B) is dominant and white (b) is recessive. What are the chances of producing white kittens when crossing a Bb male with a bb female?
    a. 0%   b. 25%   c. 50%   d. 75%   e. 100%

22. What are the chances of producing black kittens from the cross in question 21?
    a. 0%   b. 25%   c. 50%   d. 75%   e. 100%

11.4 Mendel's Laws (Page 191)

True (T) or False (F) Questions
If you believe the statement to be false, rewrite the statement as a true one.

23. When doing dihybrid crosses Mendel found the F_2 generation showed a 9:3:1 ratio of phenotypes.
    Answer: _____ Restatement: _____
    _____

24. Mendel's second law is often stated as: genes located on different chromosomes are inherited independently of each other.
    Answer: _____ Restatement: _____
    _____

11.5 How Genes Influence Traits (Pages 192–193)

25. Genes determine phenotype by specifying _____ sequence.

11.6 Why Some Traits Don't Show Mendelian Inheritance (Pages 194-199)

26. Epistasis is defined as _____

27. An Rh-negative mother pregnant with an Rh-positive child in her first pregnancy will
    a. produce "anti-Rh" antibodies that will cause clotting of her fetus's blood cells.
    b. produce "anti-Rh" antibodies that will cause clotting of any future Rh-positive fetuses' blood cells.
    c. experience a life-threatening allergic response.
    d. develop tolerance to the Rh-antigen.
    e. be required to take medication throughout her pregnancy to prevent clotting of her blood cells.

True (T) or False (F) Questions
If you believe the statement to be false, rewrite the statement as a true one.

28. In polygeny, one gene affects many traits.
    Answer: _____ Restatement: _____
    _____

29. In pleiotrophy, one gene affects many traits.
    Answer: _____ Restatement: _____
    _____

11.7   Chromosomes Are the Vehicles of Mendelian Inheritance (Pages 200-201)

30. The significance of Morgan's experiments was that
    a. it presented the first clear evidence that genes determining Mendelian traits do reside on chromosomes.
    b. it illustrated the effects of polygenic traits on the inheritance of sex-linked traits.
    c. chromosomal reduction through meiosis was proven.
    d. a method to control economically important fruit flies resulted.
    e. Both a and d.

11.8   Human Chromosomes (Pages 202-203)

31. With few exceptions trisomies of _____ are fatal.
32. XO individuals have a condition known as _____ while those with an XXY karyotype have a condition known as _____.

11.9   The Role of Mutations in Human Heredity (Pages 204-207)

33. _____ Sickle-cell anemia   a. Poor blood circulation due to abnormal hemoglobin molecules.
34. _____ Cystic fibrosis   b. Brain fails to develop in infancy.
35. _____ Hemophilia   c. Failure of blood to clot.
36. _____ Phenylketonuria   d. Failure of chloride ion transport mechanism.

37. Hemophilia involving a lack of blood protein factors VIII and IX is
    a. a dominant trait that usually only develops in individuals in their thirties.
    b. a recessive trait inherited on autosomes.
    c. a sex-linked trait.
    d. common in royal families due to several mutations in their genetic material.
    e. a dominant trait inherited on autosomes.

38. Hemoglobin in the red blood cells of individuals with sickle-cell anemia differ in _____ of the 574 amino acids.

39. Individuals who are _____ for sickle-cell anemia have increased resistance to the important infectious disease _____.

40. Tay-Sachs disease is rare in most human populations but is more common in some populations of _____ where approximately 1 in _____ infants have the disease.

41. Individuals who are heterozygous for Huntington's disease will eventually express symptoms and die because the disorder is caused by a _____.

11.10   Genetic Counseling and Therapy (Pages 208-209)

42. _____ Ultrasound
43. _____ Amniocentesis
44. _____ Chorionic villi sampling

a. Cells are removed and grown in culture.
b. Technique that can be used as early as eight weeks in pregnancy and yields rapid results.
c. Used to examine fetus for major abnormalities.

Chapter Test

1. Garden peas were a good subject for Mendel's genetic studies because
   a. they are available in many varieties.
   b. they are small, easy to grow plants.
   c. it is easy to control their fertilization.
   d. they mature quickly.
   e. all of the above.

2. To conduct genetic experiments it is necessary to begin with subjects that are
   a. haploid.
   b. heterozygous.
   c. true-breeding.
   d. easy to count.
   e. previously unstudied.

3. During your plant research project, you cross a blue-flowered variety with a yellow-flowered variety. The F_1 generation consists of all yellow-flowered plants. The P generation yellow-flowered plants are probably
   a. recessive.   b. dominant.   c. incompletely dominant.   d. sterile.   e. hybrid.

4. In the experiment in question 3 the blue-flowered plants are probably
   a. recessive.   b. dominant.   c. incomplete dominant.   d. sterile.   e. hybrids.

5. In the experiment in question 3, what would you expect to see in the F_2 generation?
   a.   all yellow-flowered plants
   b.   all blue-flowered plants
   c.   yellow-flowered and blue-flowered plants
   d.   plants with an intermediate color between yellow and blue
   e.   none of the above

6. Which of the following represents an individual that is homozygous for the dominant trait?
   a. *bb*   b. *Bb*   c. *BB*   d. *aa*   e. *Aa*

7. Which of the following represents an individual that is heterozygous for the dominant trait?
   a. *bb*   b. *Bb*   c. *BB*   d. *aa*   e. *AA*

8. Two common coat colors seen in Shetland sheepdogs (shelties) are sable and tricolor. Sable dogs (B) are blonde and tricolored dogs (b) are black with a white collar and tan eyebrows. Sable is the dominant color; tricolor is recessive. A dog that is heterozygous for coat color (Bb) is called a tri-factored sable and is darker than a dog that is homozygous for sable (BB). What are the possible coat colors in a litter that results from a cross between a homozygous sable and a tri-factored sable?
   a. all sable puppies
   b. all tricolored puppies
   c. both sable and tricolored puppies
   d. sable and tri-factored sable puppies
   e. all tri-factored sable puppies

9. What coat colors are possible from a cross between a sable and a tricolored dog?
   a. all sable puppies
   b. all tricolored puppies
   c. both sable and tricolored puppies
   d. sable and tri-factored sable puppies
   e. all tri-factored sable puppies

10. A tri-factored sheltie has a litter of puppies. Two of the pups are tricolored and the other two are tri-factored. What color was the father of the puppies?
    a. sable   b. tricolored   c. tri-factored   d. either a or b   e. either b or c

11. If the tri-factored sheltie had one sable puppy and two tri-factored puppies, and one tricolor then what color would the father have been?
    a. sable   b. tricolored   c. tri-factored   d. either a or b   e. either b or c

12. A cross between two individuals results in a ratio of 9:3:3:1 of four possible phenotypes. This is an example of a(n)
    a. monohybrid cross.   b. dihybrid cross.   c. testcross.
    d. incomplete dominance.   e. none of the above.

13. When a trait is controlled by the interaction of the products of two genes it is referred to as
    a. dominance.   d. incomplete dominance.
    b. pleiotropy.   e. aneuploidy.
    c. epistasis.

14. The tri-factored shelties mentioned earlier are an example of
    a. dominance.   d. incomplete dominance.
    b. pleiotropy.   e. aneuploidy.
    c. epistasis.

15. A DNA molecule that is complexed with proteins into a rodlike structure is called a(n)
    a. chromophore.   b. chromosome.   c. genophore.   d. genome.   e. genosome.

16. The end result of meiosis in human males is
    a. two diploid sperm cells.   d. four haploid sperm cells.
    b. four diploid sperm cells.   e. none of the above.
    c. two haploid sperm cells.

17. _____ is (are) a significant source of genetic recombination during gamete production.
    a. Mutation   b. Crossing-over   c. Nondisjunction   d. Controlled assortment   e. Both a and c

18. Which of the following is a result of nondisjunction of chromosomes during gamete formation?
    a. cystic fibrosis   b. Down syndrome   c. Turner's syndrome   d. sickle-cell anemia   e. both b and c

19. A Barr body is a(n)
    a. inactivated X chromosome.
    b. inactivated Y chromosome.
    c. abnormality seen in the cells of persons with Down syndrome.
    d. chromosome 21 trisomy complex.
    e. none of the above.

Additional Study Help

Visit the ARIS (Assessment, Review, and Instruction System) site at aris.mhhe.com for quizzes, animations, and other study tools.

12 DNA: The Genetic Material

Key Concepts Outline

A series of experiments contributed to the understanding that DNA is the molecule of heredity. (pages 214-215)
- Experiments by Avery and his coworkers and the team of Hershey and Chase showed that genes are made of DNA and not protein.

Watson and Crick described the structure of DNA as a double helix composed of two strands of nucleotides held together by hydrogen bonds between the bases. (pages 216-217)
- Chargaff's rule that A = T and G = C assisted Watson and Crick with their construction of a DNA model.
- Franklin's research helped Watson and Crick determine that DNA is in the shape of a helix.

Semiconservative replication of DNA occurs during interphase and results in two copies of DNA. (pages 218-221)
- DNA replication is facilitated by helicase, DNA polymerase, and DNA ligase.

A mutation is a change in the hereditary message. (pages 222-224)
- Mutations that change one or only a few nucleotides are called point mutations.
- They may arise as the result of errors in pairing during DNA replication, ultraviolet radiation, or chemical mutagens.
- A variety of chemical mutagens have been implicated in causing cancer.

Key Terms Matching

1. _____ Transformation
2. _____ Nucleotides
3. _____ Double helix
4. _____ Base pairs
5. _____ Complementarity
6. _____ Replication
7. _____ Helicase
8. _____ Replication fork
9. _____ DNA ligase
10. _____ Deletion
11. _____ Transposition
12. _____ Mutation
13. _____ Carcinogen

a. This allows DNA to copy itself during cell division.
b. The language of DNA.
c. Some chemicals, ultraviolet light, mutations, can act as these.
d. How genetic information was transferred to *S. pneumoniae* in Griffith's work.
e. Where the the parent DNA molecule becomes unzipped.
f. Enzyme that unwinds the double helix.
g. Loss of one or a few base pairs in a coding sequence.
h. Movement of individual genes from one place in the genome to another.
i. Enzyme that links sections of newly formed DNA together.
j. An error in a DNA molecule.
k. Making copies of a DNA molecule.
l. A double stranded molecule of DNA is called this.
m. Hydrogen bonds keep these together.

12.1   The Griffith Experiment (Page 214)

14. When Griffith injected mice with the polysaccharide coat produced by the virulent strain of *S. pneumoniae* they _____ die and when he injected them with the heat-killed *S. pneumoniae* bacterium the mice _____ die.
15. The movement of a gene from one organism to another is called _____.

12.2   The Avery and Hershey-Chase Experiments (Page 215)

16. Despite the removal of all _____ from the mixture of live R and dead S *S. pneumoniae* made by Avery and his colleagues in their experiments, transformation still took place.
17. When the _____ was destroyed in the mixture of R and S *S. pneumoniae,* all transforming activity stopped.

True (T) or False (F) Questions
If you believe the statement to be false, rewrite the statement as a true one.

18. The Hershey-Chase team of researchers used viruses to prove DNA was the genetic material causing the transformation observed in Griffith's and Avery's experiments.
    Answer: _____ Restatement: _____
    _____

19. Hershey and Chase found that the hereditary material in bacteriophages was a protein/DNA complex.
    Answer: _____ Restatement: _____
    _____

12.3   Discovering the Structure of DNA (Pages 216–217)

20. Adenine and guanine have nitrogen-containing bases with _____ ring(s) while thymine and cytosine have nitrogen-containing bases with _____ ring(s).
    a. two carbon, one carbon   d. one carbon, two carbon
    b. one carbon, one carbon   e. three carbon, two carbon
    c. two carbon, two carbon

21. X-ray diffraction experiments conducted by _____ led to the determination that DNA was a helical molecule.
    a. Francis Crick   d. Rosalind Franklin
    b. James Watson   e. Martha Chase
    c. Heinz Fraenkel-Conrat

12.4   How the DNA Molecule Copies Itself (Pages 218–222)

22. _____ refers to the need of each DNA strand to replicate one new strand of DNA when it reproduces.
23. The Meselson-Stahl experiments used _____ to determine how DNA molecules were replicated.
24. The site at which the DNA molecule begins to "unzip" is called a(n)_____.

12.5   Mutation (Pages 222–224)

25. Mutations are only inherited if they occur _____.

26. Mutations that throw the reading of the gene out of register are _____ mutations.
    a. point   b. transpositions   c. frame-shift   d. chromosomal rearrangements   e. insignificant

27. Mutations involving the alteration of one or a few base pairs in the coding sequence are _____ mutations.
    a. point   b. transpositions   c. frame-shift   d. chromosomal rearrangements   e. insignificant

28. Diesel exhaust, pesticides, and cigarette tar have been implicated in _____ cancer.
    a. breast   b. skin   c. nose   d. lung   e. bladder

Chapter Test

1. The material that transformed harmless bacteria into virulent bacteria in Avery's experiment was
   a. DNA.   b. mRNA.   c. tRNA.   d. protein.   e. RNA polymerase.

2. What is the significance of Chargaff's rule in the DNA molecule?
   a. The pairing of purines and pyrimidines in DNA results in an excess of purine in some molecules.
   b. Chargaff's rule says that a double helix should not form except at pH 7.
   c. The rules of base pairing proposed by Chargaff result in a molecule with a constant thickness.
   d. both a and b.
   e. none of the above.

3. The replication of DNA molecules in a eukaryotic cell takes place
   a. just prior to metaphase in mitosis.
   b. prior to cell division.
   c. continuously.
   d. only during meiosis.
   e. shortly after telophase.

4. What is the base sequence of a DNA strand that reads CCATAGTTTCA?
   a. CCATAGTTTCA
   b. GGATAGCCCAT
   c. GGTATCAAAGT
   d. GGTATCCGCAT
   e. none of the above

5. If a piece of DNA is found to have 430 total nucleotides and 90 of them have guanine bases, how many thymine bases are there in that piece of DNA?
   a. 250
   b. 90
   c. 340
   d. 125
   e. 180

6. _____ is the enzyme that oversees DNA replication by reading each strand and adding the appropriate next base to create a complimentary strand.
   a. DNA ligase
   b. RNA polymerase
   c. maltase
   d. helicase
   e. DNA polymerase

7. Hershey and Chase concluded that DNA, not protein, is the molecule of heredity after they observed radioactive _____ inside bacteria infected by viruses.
   a. carbon
   b. phosphorus
   c. nitrogen
   d. sulfur
   e. oxygen

8. The two strands of nucleotides that comprise DNA are connected by ____ between the bases.
   a. hydrogen bonds
   b. helicase
   c. fibrin
   d. ionic bonds
   e. triple covalent bonds

9. Meselson and Stahl's results of one heavy/light hybrid DNA molecule in the F_1 generation and one unlabeled molecule and one heavy/light hybrid molecule in the F_2 generation supports the theory of
   a. dispersive replication.
   b. semiconservative replication.
   c. conservative replication.

10. Chemical or physical damage to DNA can cause _____ mutations involving one or a few nucleotides.
    a. single   b. simple   c. insignificant   d. point   e. patterned

Additional Study Help

Visit the ARIS (Assessment, Review, and Instruction System) site at aris.mhhe.com for quizzes, animations, and other study tools.

13 How Genes Work

Key Concepts Outline

DNA serves as a template on which mRNA is assembled in a process called transcription. (page 228)
- The enzyme RNA polymerase binds the DNA at a site called a promoter.
- As RNA polymerase moves along the strand of DNA being transcribed, each DNA nucleotide is paired with its complimentary RNA nucleotide (A with U, G with C) which builds an mRNA strand.

The process of reading mRNA and turning its information into a sequence of amino acids associated with a certain protein is called translation. (pages 229-231)
- The genetic information is encoded in three-nucleotide units called codons that correspond to certain amino acids.
- Translation takes place at ribosomes that move along the mRNA, adding an amino acid to the end of a growing protein chain as it passes each codon. Transfer RNA delivers amino acids to ribosomes.

Genes are uninterrupted stretches of DNA nucleotides in prokaryotes; in eukaryotes genes are fragmented. (pages 232-233)
- In eukaryotes the sequence of nucleotides that encodes an amino acid sequence for a protein is called an exon. Exons are broken up by extra noncoding DNA nucleotides called introns.
- There are multiple copies of many eukaryotic genes scattered around on different chromosomes.

Most genes are regulated by controlling the rate at which they are transcribed. (pages 234-235)
- Bacteria shut off genes by attaching a repressor protein to or near the promoter, so the polymerase is unable to bind.
- Eukaryotes can use many regulatory proteins simultaneously.

Key Terms Matching

1. _____ Transcription
2. _____ mRNA
3. _____ RNA polymerase
4. _____ Genetic code
5. _____ Codon
6. _____ Translation
7. _____ Ribosome
8. _____ tRNA
9. _____ Anticodon
10. _____ Promoter
11. _____ Repressor
12. _____ Operon
13. _____ Enhancer
14. _____ Intron

a. This is made from a DNA template.
b. Made of three nucleotides, specifies an amino acid.
c. Where RNA polymerase binds to the DNA molecule.
d. Using mRNA to direct production of a protein molecule.
e. A set of instructions for making an organism.
f. A three-nucleotide sequence found on a tRNA molecule.
g. The "extra stuff" in a DNA molecule.
h. Made of protein and rRNA, consists of 2 subunits.
i. When bound to the operator it blocks movement of polymerase to the gene.
j. To make mRNA from a DNA template.
k. Enzyme that transcribes DNA.
l. Delivers amino acids to the growing polypeptide chain.
m. A cluster of genes transcribed as a unit.
n. Located upstream from the gene it regulates.

13.1   Transcription (Page 228)

True (T) or False (F) Questions
If you believe the statement to be false, rewrite the statement as a true one.

15. mRNA is the product of DNA transcription.
    Answer: _____ Restatement: _____

_____

16. RNA polymerase binds to one strand of a DNA molecule at a site called the promoter.
Answer: _____ Restatement: _____

_____

17. A codon may specify a(n) _____ or be translated as _____.

18. The genetic code is _____ for bacteria, plants, eagles, and humans.

13.2   Translation (Pages 229-231)

19. During translation the mRNA molecule
    a. sits between the the tRNA molecule and the large subunit of the ribosome.
    b. moves between the two subunits of the ribosome.
    c. brings amino acids from the cytoplasm to the tRNA molecule bound to the ribosome.
    d. binds amino acids together in the growing polypeptide chain.
    e. is not involved.

20. tRNA molecules with bound amino acids deliver them to the
    a. P site on the ribosome.
    b. 3′ end of the polypeptide chain.
    c. 5′ end of the polypeptide chain.
    d. nucleus.
    e. A site on the ribosome.

13.3   Architecture of the Gene (Page 232-233)

21. Clusters of almost identical genes in cells are called _____.
22. Genes that can move from one chromosome location to another are known as _____.

13.4   Turning Genes Off and On (Pages 234-235)

True (T) or False (F) Questions
If you believe the statement to be false, rewrite the statement as a true one.

23. Activators are regulatory genes that allow enhancers to bind to the promoter site on a DNA molecule.
Answer: _____ Restatement: _____

_____

24. DNA loops around so the enhancer is positioned near the promoter it regulates.
Answer: _____ Restatement: _____

_____

Chapter Test

1. What would be the base sequence of an mRNA molecule that was complimentary to a DNA strand CCATAGTTTCA?
    a. GGTATCAAAGT   d. GGUAUACAAGU
    b. CCATAGTTTCA   e. both a and b
    c. CCUAUGAAAGU

2. The production of mRNA is called
    a. transduction.   b. transformation.   c. transcription.   d. translation.   e. promotion.

3. Which of the following is the most accurate description of RNA polymerase?
    a. RNA polymerase is used to initiate the production of tRNA from an mRNA template.
    b. RNA polymerase binds to a strand of DNA and moves along the strand to make a strand of mRNA.
    c. RNA polymerase is responsible for providing a termination signal to the DNA strand at the end of
       rRNA synthesis.
    d. DNA ligase and RNA polymerase work together to bind the nucleotides in a newly forming DNA strand.
    e. RNA polymerase joins the two DNA after translation has occurred.

4. During transcription
    a. both DNA strands serve as a template for the synthesis of mRNA.
    b. the two DNA strands alternate as the template for the synthesis of tRNA.
    c. one of the DNA strands serves as a template for the synthesis of mRNA.
    d. mRNA serves as a template for the synthesis of rRNA.
    e. rRNA is assembled into complexes with proteins to make new ribosomes.

5. Each amino acid is specified by
    a. several genes.
    b. an operon and a promoter.
    .   c. a codon.
    d. markers of the surface of the cell requiring it for growth.
    e. an enhancer.

6. In ribosomes the rRNA is located
    a. on the surface of the large subunit.
    b. complexed with several proteins in the large subunit.
    c. in both the small and the large subunits.
    d. in the small subunit.
    e. none of the above.

7. Why can ribosomes bind to a sequence occurring at the beginning of a gene?
    a. Because of the matching, exposed sequence of rRNA nucleotides found on the small subunit of the ribosome.
    b. Because of the effects of certain hormones that are associated with DNA synthesis.
    c. Because the ribosomes' anti-codon complements the codon.
    d. Both a and b.
    e. None of the above.

8. What part of the tRNA molecule binds to mRNA?
    a. The single-stranded nucleotide "tail" of the tRNA molecule binds to a specific codon on the mRNA molecule.
    b. The anticodon loop on the tRNA molecule binds to a specific codon on the mRNA molecule.
    c. The tRNA codon binds to the anticodon on the mRNA molecule.
    d. The small subunit of the tRNA binds to the promoter on the mRNA molecule.
    e. The large subunit of the tRNA molecule binds to the anticodon on the mRNA molecule.

9. What happens when a ribosome reaches a codon on the mRNA molecule that does not specify any of the 64
    tRNA anticodons?
    a. The ribosome begins to read the mRNA molecule in the opposite direction.
    b. The ribosome releases the newly formed protein molecule.
    c. The ribosome subunits separate.
    d. The newly formed protein begins to break up into smaller molecules.
    e. Both b and c.

10. Which of the following mechanisms would function to prevent transcription?
    a. the addition of excess substrate to the environment
    b. binding of a repressor to the operator site on the gene
    c. binding of an enhancer to the repressor site on the gene
    d. changing the pH of the environment
    e. both a and d

11. What effect would the addition of lactose have on a repressed *lac* operon?
    a. The operator site on the operon would move.
    b. It would reinforce the repression of that gene.
    c. It would result in the repression of other genes in that area.
    d. The *lac* operon would be transcribed.
    e. It would have no effect whatsoever.

12. After transcription, the mRNA molecule must _____ before it is translated into a protein molecule.
    a. assume the correct three-dimensional sequence
    b. adjust to its new environment
    c. have its introns removed
    d. have its exons removed
    e. complex with a tRNA molecule

Additional Study Help

Visit the ARIS (Assessment, Review, and Instruction System) site at aris.mhhe.com for quizzes, animations, and other study tools.

14 The New Biology

Key Concepts Outline

Entire genomes have recently been revealed. (pages 242-243)
- Automated DNA sequencing technology has facilitated the study of genomics.

Sequencing of the human genome was completed in 2000. (pages 244-245)
- The human genome contains 3.2 billion base pairs.
- Very little of the human genome is comprised of DNA that encodes proteins.

DNA can be cut by restriction enzymes into fragments with "sticky" ends. (pages 246-248)
- The fragments of DNA can then be inserted into other DNA molecules.

Genetic engineering has facilitated advances in the fields of medicine and agriculture. (pages 249-255)
- Medically important proteins such as insulin, human growth factors, and anticoagulants and harmless vaccines have been developed.
- Plants have been genetically engineered to be disease and herbicide resistant, hardier and more nutritious.

Successful cloning of farm animals has been done. (pages 256-259)
- A lack of genomic imprinting often causes the cloning of farm animals to fail.

Damaged or lost tissues may be replaced by cloned tissues. (pages 260-263)
- Embryonic stem cells may be used to cure diabetes, Parkinson's disease and damaged muscle or nerve tissue.

Hereditary diseases like cystic fibrosis may some day be cured by transferring a healthy gene into the cells of affected tissues. (pages 265-266)
- Early attempts to cure cystic fibrosis with gene therapy were unsuccessful, but new vectors may make curing hereditary diseases possible.

Key Terms Matching

1. _____ Genome
2. _____ Single copy genes
3. _____ Multigene families
4. _____ Transposable elements
5. _____ Restriction enzyme
6. _____ DNA library
7. _____ Vector
8. _____ Piggyback vaccine
9. _____ Genomic imprinting
10. _____ Totipotent
11. _____ Adult stem cell
12. _____ Gene transfer therapy

a. Produce recessive Mendelian inheritance.
b. Introduction of healthy genes into cells lacking them.
c. Made from a modified and still harmless virus.
d. Forms specific major tissue.
e. Clusters of related but distinctly different genes.
f. Has the ability to form any body tissue or an adult animal.
g. Blocks a cell's ability to read certain genes.
h. Mobile bits of DNA.
i. Used to cut DNA into fragments with sticky ends.
j. Genes and other DNA of an organism.
k. Collection of DNA fragments of an organism's DNA.
l. Vehicle that carries a gene into a cell.

14.1   Genomics (Pages 242-243)

13. The development of _____ has made sequencing of genomes practical in the past several years.
    a. genomes   b. DNA sequencing machines   c. DNA microassays   d. biochips

14. All of an organism's genes are referred to as its
    a. DNA complement.   b. genomic.   c. genome.   d. chromosomes.   e. Both a and c are correct.

15. The first eukaryotic genome to be sequenced was that of
    a. *E. coli.*   b. *Arabidopsis.*   c. *Drosophila.*   d. *C. elegans.*   e. bacterial virus ΦX174.

14.2   The Human Genome (Pages 244–245)

True (T) or False (F) Questions

If you believe the statement to be false, rewrite the statement as a true one.

16. Noncoding segments of DNA called exons are separated by shorter coding regions called introns.
    Answer: _____ Restatement: _____
    _____

17. Multigene families contain far fewer genes than do tandem gene clusters.
    Answer: _____ Restatement: _____
    _____

18. Transposable elements were discovered by Barbara McClintock and are tiny bits of DNA that can move from one location to another on a chromosome.
    Answer: _____ Restatement: _____
    _____

19. Approximately 90% of DNA in a genome is actually coded into functional proteins.
    Answer: _____ Restatement: _____
    _____

14.3   A Scientific Revolution (Pages 246-248)

20. In genetic engineering, genes can be
    a. moved from one organism to potentially any other organism.
    b. easily converted to produce any product we desire.
    c. turned off or on at our whim.
    d. used to cure all known diseases.
    e. used to create plants resistant to all insect pests.

21. Sticky ends are the result of
    a. treatment of a nucleotide sequence with DNA ligase.
    b. exposure of eukaryotic DNA to a prokaryotic plasmid.
    c. infection of a cell with a bacteriophage.
    d. DNA breaking down in the presence of reverse transcriptase.
    e. cutting by restriction enzymes "off center" in a specific nucleotide sequence.

22. In order to produce a DNA fragment that can be inserted into the DNA of a second organism one would need to
    a. create a compatible DNA segment from an mRNA template.
    b. find a virus that could serve as a vector for transfer of the donor DNA.
    c. make sure both organisms have compatible DNA.
    d. be sure both donor and recipient DNA were prokaryotic or eukaryotic.
    e. cut the DNA of the donor and recipient cells with the same restriction enzyme.

14.4   Genetic Engineering and Medicine (Pages 249–250)

23. _____	Interleukins	a. Stimulates white cell production.
24. _____	Erythropoietin	b. Activates white blood cells. Used to treat HIV and cancer.
25. _____	Interferons	c. Halts the production of viruses.
26. _____	Colony-stimulating factors	d. Used to treat individuals with kidney disorders.

14.5   Genetic Engineering and Agriculture (Pages 251–255)

27. List three ways in which genetic modification has had a positive impact on crop plants.

28. Glyphosate prevents plants from growing by
   a. cleaving critical DNA in the plant and halting cell division.
   b. preventing the production of gametes.
   c. preventing the formation of spindle fibers and halting mitosis.
   d. stopping them from synthesizing proteins.
   e. both a and d.

14.6   Reproductive Cloning (Pages 256-259)

29. _____ cells were combined with egg cells to successfully clone Dolly.
30. In cloning the sheep Dolly, a(n) _____ _____ triggered cell division.
31. Discuss why cloning often fails.

14.7   Embryonic Stem Cells (Pages 260-261)

32. Cells that have the ability to form any body tissue are known as _____.
33. Where are embryonic stem cells found?

14.8   Therapeutic Cloning (Pages 262–263)

34. Discuss problems that need to be overcome in therapeutic cloning.

14.9   Gene Therapy (Pages 265-266)

True (T) or False (F) Question
If you believe the statement to be false, rewrite the statement as a true one.

35. The AAV virus does not elicit a strong immune response.
   Answer: _____ Restatement: _____

Chapter Test

1. Transposable elements occur at a rate of about _____% in the human genome.
   a. 1   b. 24   c. 20   d. 3   e. 45

2. Tandem clusters of DNA affect the production of a gene product in which of the following ways?
   a. The gene product is more accurately produced than if it was transcribed from single-copy genes.
   b. The gene product is left unmodified for later use by the cell.
   c. Several products can be produced from each tandem DNA cluster.
   d. The gene product can be produced rapidly and in large amounts.
   e. Production of the gene product is blocked.

3. Recessive Mendelian traits are expressed under which of the following conditions?
   a. When they exist as single-copy genes.
   b. When they exist as segmental duplications of genes.
   c. When the organism mutates to a polyploid form.
   d. Only in plant species that undergo alternation of generations.
   e. None of the above are true.

4. Transposable segments of DNA are able to
   a. be read from either of two directions.
   b. move from one chromosome to another.
   c. take the place of damaged mRNA molecules.
   d. change depending on environmental conditions.
   e. produce alternate versions of a gene product.

5. Cuts in DNA are sealed with

    a. restriction enzymes. b. ligases. c. reverse transcriptase. d. polymerase. e. plasmids.

6. In order for plants to resist the effects of herbicides, such as Roundup, they must be

    a. monocots. d. able to synthesize glyphosate.

    b. dicots. e. resistant to glyphosate.

    c. annuals.

7. Inserting *Bacillus thyringiensis* (Bt) in crops

    a. increases glyphosate production.

    b. is harmful to humans.

    c. produces proteins toxic to pests.

    d. none of the above.

    e. a and b are correct.

8. Dolly was cloned from fusing together

    a. a mammary cell without a nucleus and an egg cell without a nucleus.

    b. a sperm cell without a nucleus and an egg without a nucleus.

    c. an egg cell without a nucleus and a mammary cell with a nucleus.

    d. none of the above.

9. Adding a healthy *cf* (cystic fibrosis) gene to adenovirus proved unsuccessful because

    a. the adenovirus mutated.

    b. the adenovirus rejected the *cf* gene.

    c. the adenovirus induced a strong immune response.

    d. none of the above.

10. The *adeno-associated virus* (AAV) is a much more promising vector because

    a. it does not elicit a strong immune response.

    b. it has only two genes.

    c. it needs adenovirus to replicate.

    d. all of the above.

Additional Study Help

Visit the ARIS (Assessment, Review, and Instruction System) site at aris.mhhe.com for quizzes, animations, and other study tools.

15   Evolution and Natural Selection

Key Concepts Outline

Microevolution leads to macroevolution. (pages 270–271)
- Adaptation to local habitats leads to divergence in the evolution of ecological races.
- Isolating mechanisms then reinforce the differences, leading to reproductive isolation and species formation.

The occurrence of evolution is supported by the fossil record, comparative anatomy and the genetic record. (pages 272-276)
- If one dates fossils, and orders them by age, progressive changes are seen.
- Embryo comparison and homologous and analogous structures support evolution.
- Comparisons of DNA and different proteins provide additional strong evidence that evolution has occurred.

Evolution is controversial because people mistakenly believe evolution challenges their religious beliefs. (pages 277-279)
- Science requires critical analysis of observations and data, and the theory of evolution is supported by critical analysis; criticisms of evolution are without scientific merit.

In a population not undergoing significant evolutionary change, two alleles present in frequencies p *and* q *will be distributed among the genotypes in the proportions* $p^2 + 2pq + q^2$, *the Hardy-Weinberg equilibrium.* (pages 282-283)
- Hardy-Weinberg equilibrium will be true only if the population size is large, random mating occurs, there are no mutations and no gene flow, and there is no natural selection.

Allele frequencies change in nature due to mutation, migration, drift, nonrandom mating, and selection. (pages 284-288)
- The founder effect and population bottleneck are the result of genetic drift.
- Selection for certain traits can favor intermediate characteristics or either extreme.

A mutation in hemoglobin causes a condition known as sickle-cell anemia. (pages 288-289)
- This recessive mutation is common in central Africa because it renders heterozygous individuals resistant to malaria.

Experiments to test hypotheses about how evolution occurs indicate that natural selection can cause evolutionary change to occur quickly. (pages 290-291)
- The number of spots seen on guppies raised in laboratory greenhouses and observed in field experiments corresponded to the level of predation of the environment.

The absence of gene flow among populations makes speciation occur more quickly. (pages 292-297)
- Reproductively isolated populations constitute different species because they do not mate or produce fertile offspring with other populations.

Key Terms Matching

1. _____ Fossil
2. _____ Radioisotopic testing
3. _____ Molecular clock
4. _____ Homologous structure
5. _____ Microevolution
6. _____ Allele frequency
7. _____ Hardy-Weinberg equilibrium
8. _____ Directional selection
9. _____ Balancing selection
10. _____ Punctuated equilibrium
11. _____ Industrial melanism
12. _____ Species
13. _____ Isolating mechanism

a. The rate at which a trait is found in the gene pool of a population.
b. An artificial way to predict evolution in a population.
c. Caused dark moths to survive while light ones disappeared.
d. A mountain or river can be this to a population.
e. A tendency for individuals in a population to have similar phenotypes.
f. Spurts of evolution.
g. Structures derived from the same structure in a common ancestor.
h. For example, the changes in cytochrome c in various species.
i. Small changes in a population of organisms.
j. Individuals at an extreme of the phenotype are favored.
k. Mineralized remains of an organism.
l. Used to date fossils.
m. Organisms with the potential to successfully interbreed in nature.

15.1   Evolution: Getting from There to Here (Pages 270–271)

14. Adaptation is the result of _____ changes that increase the likelihood of survival and _____.

15. Evolution is the end result of natural selection on _____ within populations.

16. Macroevolution refers to
    a. change on a grand scale.
    b. change in gene frequencies within populations.
    c. physical changes that occur during an organism's lifetime.
    d. none of the above.

15.2   The Evidence for Evolution (Pages 272–275)

17. Darwin's examinations of fossils relied on _____ dating to determine the evolution of species.
    a. absolute
    b. carbon
    c. use of a variety of radioactive isotopes
    d. relative
    e. none of the above

True (T) or False (F) Questions
If you believe the statement to be false, rewrite the statement as a true one.

18. If evolutionary theory is correct, every evolutionary change involves the substitution of new versions of genes for old ones.
    Answer: _____ Restatement: _____
    _____

19. When examining the evidence provided by a molecular clock, such as cytochrome c, more changes in the nucleotide sequences between two organisms means they are less closely related.
    Answer: _____ Restatement: _____
    _____

20. _____ structures are structures which have evolved from the same body part present in a common ancestor while _____ structures develop in unrelated lines but are modified by natural selection to look the same and have the same function.

15.3   Evolution's Critics (Pages 277-281)

21. Critics of evolution have recently argued that life is too complex to be the result of natural selection, so _____ should be taught in science classes as an alternative to the theory of evolution.

15.4   Genetic Change Within Populations: The Hardy-Weinberg Rule (Pages 282-283)

22. _____ Allele   a. The less common of two alleles in a population.
23. _____ p   b. Alternate versions of a gene.
24. _____ q   c. The sum of all variants in a population.
25. _____ 1   d. The more common of two alleles in a population.

15.5   Agents of Evolution (Pages 278–279)

26. _____ Artificial selection   a. Breeding the "best to the best."
27. _____ Bottleneck effect   b. This alone does not have much effect on allele frequency.
28. _____ Genetic drift   c. Most organisms in a population are killed in a flood.
29. _____ Mutation   d. The founder effect.

30. The elimination of intermediate phenotypic variants from a population illustrates _____ selection.

31. The elimination of both extremes of phenotype from a population illustrates _____ selection.

15.6 Sickle-Cell Anemia (Pages 288–289)

True (T) or False (F) Questions
If you believe the statement to be false, rewrite the statement as a true one.

32. People who are homozygous for sickle-cell anemia are less susceptible to malaria.
    Answer: _____ Restatement: _____

    _____

33. The gene for cystic fibrosis may play a role in resistance to typhoid fever.
    Answer: _____ Restatement: _____

    _____

15.7 Selection on Color in Guppies (Pages 290–291)

34. The results of experiments conducted using the guppy *Poecilia reticulata* suggest that
    a. increased predation increases the number of spots.
    b. decreased predation increases the number of spots.
    c. natural selection can lead to rapid evolutionary change.
    d. both b and c are correct

15.8 The Biological Species Concept (Page 292)

35. German shepherds and poodles are able to mate and produce viable fertile offspring (strange looking though they might be!) so they must be:
    a. members of two closely related but similar species.
    b. members of different species.
    c. genetically compatible members of the same subspecies.
    d. members of the same species.
    e. members of the same ecological race.

36. Which of the following plays a role in speciation?
    a. Lack of phenotypic variants in a population
    b. Natural selection reinforcing differences between two races
    c. Geographic isolation of organisms that were once in the same population
    d. The ability of organisms to interbreed with other members of the population
    e. Both b and c

37. What is meant by the biological species concept?

15.9 Isolating Mechanisms (Pages 294-295)

38. The males of two species of birds use different mating dances to attract their mates. This is an example of _____ isolation.
    a. postzygotic   b. behavioral   c. ecological   d. hybrid incompatibility   e. hybrid breakdown

39. One species of snake lives exclusively in the water while another species of snake inhabits trees. This is an example example of _____ isolation.
    a. postzygotic   b. behavioral   c. ecological   d. hybrid incompatibility   e. hybrid breakdown

40. _____ Geographic isolation   a. Species differ in mating rituals.
41. _____ Temporal isolation   b. Structural differences prevent mating.
42. _____ Behavioral isolation   c. Species reproduce in different seasons.
43. _____ Ecological isolation   d. Species separated by physical barriers.
44. _____ Mechanical isolation   e. Species occur in different habitats.

15.10 Working with the Biological Species Concept (Pages 296–297)

45. _____ Allopatric speciation   a. Having more than two sets of chromosomes.
46. _____ Sympatric speciation   b. Geographically isolated populations.
47. _____ Polyploidy   c. One species splits into two.

Chapter Test

1. Molecular clocks measure
   a. differences in nucleotide sequences between different species.
   b. the age of fossils more than 150 million years old.
   c. the half-life of potassium-16.
   d. both a and b.
   e. both b and c.

2. After examining the evidence related to the evolution of hemoglobin you might conclude that
   a. lamprey globin evolved prior to insect globin.
   b. plant globins evolved from an entirely different type of gene than did human hemoglobin.
   c. it is strong evidence that modern organisms evolved from simpler forms.
   d. fish are more closely related to whales than they are to kangaroos.
   e. both a and d.

3. Insect wings and the wings of bats are
   a. analogous structures.   d. structurally similar.
   b. homologous structures.   e. the result of microevolution.
   c. simple structures.

4. In the year 2317, a group of 507 pioneers colonize Mars. The incredible distance between Mars and Earth excludes the possibility of new members joining the colony. This situation is an example of
   a. genetic drift.   b. mutation.   c. the founder effect.   d. artificial selection.   e. outcrossing.

5. Which of the following conditions are necessary for the Hardy-Weinberg theorem to be valid?
   a. The population being examined must be small and have a limited gene pool.
   b. New alleles must be added to the population on a regular basis.
   c. All of the individuals in the population must be heterozygous for all traits.
   d. Natural selection is acting on the population.
   e. Random mating is occurring within the population.

6. Much of the soil in Sedona, Arizona, is brick red in color. Several birds and small mammals living in that area include ants in their diet. Although both black and red ants live there, red ants are much more common. This might be due to
   a. directional selection.   d. outcrossing.
   b. stabilizing selection.   e. heterozygote advantage.
   c. disruptive selection.

7. Two species of birds that do not mate because they respond to different mating dances are separated by
   a. geographical isolation.   d. a postzygotic isolating mechanism.
   b. a prezygotic isolating mechanism.   e. both b and c.
   c. behavioral isolation.

8. Profound changes, such as major episodes of extinction, are an example of
   a. punctuated equilibrium.   d. selective stabilization.
   b. macroevolution.   e. directional stabilization.
   c. microevolution.

9. If the punctuated equilibrium model is correct, then
    a. many evolutionary changes occur in a short period of time which are followed by periods of little evolutionary change.
    b. only eukaryotic, nonphotosynthetic species are subject to natural selection.
    c. Darwin's finches evolved by this mechanism.
    d. evolution proceeds with gradual, successive changes in a given evolutionary line.
    e. it is responsible for the many races of humans existing on our planet today.

10. Although there are gaps in the fossil record
    a. the evolution of prokaryotes is clearly outlined.
    b. it provides a clear record of continuous evolutionary change.
    c. it is not accepted as proof of evolution by modern scientists.
    d. most mammalian species are clearly shown to have evolved from fungal-like eukaryotic organisms.
    e. none of the above are true.

11. The embryos of numerous vertebrate animals
    a. develop at remarkably similar rates.
    b. are exactly alike until the 13th week of development.
    c. share many primitive features early in their development.
    d. provided much of the evidence for Darwin's argument in favor of evolution.
    e. both a and b are true.

12. In cats, if B represents black and is dominant and b represents tan and is recessive, what are the chances of tan kittens being produced by mating a BB cat to a Bb cat?
    a. 0%   b. 25%   c. 50%   d. 75%   e. 100%

13. In cats, if B represents black and is dominant and b represents tan and is recessive, what are the chances of black kittens being produced by mating a BB cat to a Bb cat?
    a. 0%   b. 25%   c. 50%   d. 75%   e. 100%

14. Genetic drift
    a. has a significant effect on large, diverse populations.
    b. only plays a role in the evolution of animals.
    c. usually only occurs in small populations.
    d. is the ultimate source of variation.
    e. is the result of profound mutation.

15. When disruptive selection occurs in a population, individuals
    a. in the middle range of the phenotype become more common.
    b. in the middle range of the phenotype always become extinct.
    c. in the middle range of the phenotype are selected against.
    d. with an extreme form of the phenotype further diversify.
    e. tend to migrate to a new location.

Additional Study Help

Visit the ARIS (Assessment, Review, and Instruction System) site at aris.mhhe.com for quizzes, animations, and other study tools.

16   Exploring Biological Diversity

Key Concepts Outline

Linnaeus invented the binomial system for naming species, a vast improvement over complex polynomial names. (page 302)
- Linnaeus used a two-part name for different species and also grouped similar organisms into higher-level categories based on similar characteristics.

An organism's scientific name is its genus and species, which are given in Latin. (page 303)
- The genus of an organism is the first word of its scientific name and the species of an organism is the second. The genus name is capitalized, but the species name is not, and both are written in italics.

The Linnaean system of classification is a hierarchical system in which higher categories are more general groupings of organisms. (page 304)
- A domain is the broadest level of classification and is followed by kingdom, phylum, class, order, family, genus and species.

The "biological species concept" defines species as groups that cannot successfully interbreed. (page 305)
- This concept works well for animals and outcrossing plants but poorly for other organisms.

Taxonomy is of great practical importance. (pages 306-308)
- Systematics is the study of the evolutionary relationships among a group of organisms.
- Cladistics builds family trees by clustering together those groups that share an ancestral "derived" character.
- Phylogenetic systematics classifies organisms based on a large amount of information available, giving due weight to the evolutionary significance of certain characters.

Living organisms are grouped into three domains, the largest and most general groupings of organisms. (pages 310-314)
- Two domains, Bacteria and Archaea, contain the prokaryotes.
- The Eukarya domain is divided into four kingdoms: Protista, Fungi, Plantae, and Animalia.
- Endosymbiosis is the means by which eukaryotic cells acquired mitochondria and chloroplasts.

Key Terms Matching

1. _____ Classification
2. _____ Genus
3. _____ Binomial system
4. _____ Taxonomy
5. _____ Biological species concept
6. _____ Systematics
7. _____ Phylogenetic trees
8. _____ Cladistics
9. _____ Derived character
10. _____ Outgroup
11. _____ Domain

a. The study of groups.
b. A taxonomic level higher than kingdom.
c. Similarities derived from a common ancestor.
d. A group of related species.
e. Multilevel grouping of organisms.
f. Genus and species names for each organism.
g. A group of potentially interbreeding organisms.
h. A "different" organism that serves as a baseline for comparison.
i. Illustrates relationships between organisms.
j. A branching diagram representing phylogeny.
k. The study of evolutionary trees.

16.1   The Invention of the Linnaean System (Page 302)

12. Prior to the development of the Linnaean system, Aristotle developed a system classifying animals as:
    a. _____ b. _____ or c. _____.
13. A _____ system developed in the mid-1700s used a string of Latin words and phrases to name organisms.

16.2   Species Names (Page 303)

True (T) or False (F) Questions
If you believe the statement to be false, rewrite the statement as a true one.

14. Common names are useful in describing organisms when traveling to other parts of the world.
    Answer: _____ Restatement: _____
    _____

15. The first part of a binomial name refers to the genus while the second name is the species name.
    Answer: _____ Restatement: _____
    _____

16.3   Higher Categories (Page 304)

16. Members of the same phylum are also members of the same
    a. kingdom.   b. class.   c. order.   d. family.   e. genus.

17. Members of which of the following would be most likely to be able to interbreed and produce a hybrid?
    a. kingdom   b. class   c. order   d. family   e. genus

16.4   What Is a Species? (Pages 305)

18. According to Ray's species definition horses and donkeys are
    a. members of the same species because they can produce viable offspring.
    b. members of the same species because they are similar in appearance.
    c. not members of the same species because their offspring are sterile.
    d. not members of the same species because they are different sizes.
    e. in the same species as mules.

True (T) or False (F) Question
If you believe the statement to be false, rewrite the statement as a true one.

19. The biological species concept is controversial and not widely applicable to plants.
    Answer: _____ Restatement: _____
    _____

16.5   How to Build a Family Tree (Pages 306–302)

20. A(n) _____ is a group of organisms related by descent.
21. The evolutionary history of an organism is referred to as its _____.
22. Organisms that are lower on a cladogram evolved _____ than those at the top.

16.6   The Kingdoms of Life (Page 310)

23. _____ Archaebacteria   a. Photosynthetic multicellular organisms.
24. _____ Eubacteria   b. Eukaryotic and multicellular; ingest their nutrients.
25. _____ Protista   c. Your "garden variety" bacteria.
26. _____ Fungi   d. Molds and mushrooms.
27. _____ Plantae   e. The methanogens are members of this kingdom.
28. _____ Animalia   f. Unicellular and eukaryotic; for example, amoeba.

16.7   Domain Bacteria (Page 311)

29. _____ has/have provided evidence that the eubacteria are more closely related to the eukaryotes than to the archaebacteria.
    a. Fossils   b. RNA sequencing   c. Cytochrome c comparison   d. Cladistics   e. Both a and d

30. Chitin is a polysaccharide that is a cell wall component of
    a. bacteria.   b. plants.   c. fungi.   d. protozoans.   e. animals.

16.8   Domain Archaea (Page 312)

31. The archaebacteria are grouped into three categories:
    a. _____   b. _____   c. _____.

32. The halophiles have an extreme tolerance for _____ and have unique pigment called
    _____.

16.9   Domain Eukarya (Pages 313–314)

33. _____ are believed to be the precursors to mitochondria.
    a. Cyanobacteria   b. Purple bacteria   c. The cellular slime molds   d. Fungi   e. None of the above

Chapter Test

1. Arranging organisms into a multilevel system is called
    a. classification.   b. nomenclature.   c. speciation.   d. paleontology.   e. ornithology.

2. In the name *Escherichia coli, coli* is the _____ name.
    a. genus   b. species   c. family   d. common   e. phylum

3. The difficulty with polynomial names was that they
    a. were not descriptive enough.   d. were long and confusing.
    b. were not consistent.   e. all of the above.
    c. could only be applied to animals.

4. Taxonomy
    a. is closely related to classification.   d. both a and b.
    b. identifies and names taxa of organisms.   e. both b and c.
    c. names specific organisms.

5. Which of the following is the correct way to write a scientific name?
    a. Canis familiaris   d. <u>Canis familiaris</u>
    b. Canis Familiaris   e. *Canis familiaris*
    c. canis familiaris

6. Which of the following is the correct order of taxonomic categories beginning with the largest and ending with the smallest group?
    a. phylum, kingdom, class, family, genus, order, species
    b. kingdom, phylum, class, order, family, genus, species
    c. class, family, species, order, phylum, kingdom, genus
    d. species, genus, family, order, class, phylum, kingdom
    e. none of the above

7. Organisms that are members of the same class are also members of the same
    a. species.   b. family.   c. genus.   d. phylum.   e. both b and d

8. While examining water samples that you collected at a local pond, you observe a unicellular organism that has a nucleus and organelles. That organism is a member of the kingdom
    a. Eubacteria.   b. Protista.   c. Animalia.   d. Plantae.   e. Fungi.

9. One limitation of the biological species concept is that it
   a. does not apply to asexually reproducing organisms.
   b. is only applicable to plant species.
   c. is only applicable to animal species.
   d. is outdated.
   e. all of the above.

10. While conducting research in the rain forest you collect a number of beetles. Several of them have similar morphologies but different coloring. How could you determine if they are members of the same species?
    a. by determining if they produce viable, fertile offspring
    b. by determining if they live in the same niche
    c. by determining if they eat the same foods
    d. by further examination of their morphologies for significant differences
    e. all of the above.

11. It is later determined that the beetles mentioned in question 10 are able to produce viable, fertile offspring. The different marking could mean that they are
    a.   morphologically isolated.   b. ecotypes.   c. ecological races.   d. serovars.   e. both b and c

12. Phylogeny
    a. refers to the evolutionary relationships between organisms.
    b. provides information about community interactions.
    c. explains the physiology of organisms.
    d. is only valid in relation to sexually reproducing organisms.
    e. both a and b.

13. Which of the following is a difference between cladistics and phylogenetic systematics?
    a. Phylogenetic systematics reflects relative relationships between organisms.
    b. Cladistics relies more heavily on morphological characteristics than physiological characteristics.
    c. The cladistic approach provides a more accurate picture of the evolutionary relationships between organisms.
    d. Phylogenetic systematics is the better approach when information about the organisms being studied is plentiful.
    e. both a and b.

14. Although dogs and coyotes can produce fertile offspring, they are not members of the same species.
    Why is this?
    a. Differences in morphology.
    b. Differences in their behaviors.
    c. Many of their offspring are sterile.
    d. They inhabit different environments.
    e. They require different amino acids in their diets.

15. Although two birds living in the same forest share many similarities, they rarely interbreed. One species lives in the upper branches of trees while the other lives in low-growing shrubs. Their different niches act as a(n)
    a. separating mechanism.   d. cladistic barrier.
    b. isolating mechanism.   e. none of the above.
    c. species isolator.

Additional Study Help

Visit the ARIS (Assessment, Review, and Instruction System) site at aris.mhhe.com for quizzes, animations, and other study tools.

17   The First Single-Celled Creatures

Key Concepts Outline

Life appeared on earth 2.5 billion years ago. (pages 318–319)
- In principle, there are three possible origins of living things: extraterrestrial, special creation, and evolution.
- The Miller-Urey experiment and bubble model offer explanations of how life may have arisen spontaneously.

Prokaryotes are the most ancient and simplest form of life on earth. (pages 321–323)
- Bacteria and archaea are small, simply organized, single cells that lack an organized nucleus.
- Prokaryotes reproduce by binary fission.
- Some bacteria gain genetic diversity by transferring plasmids in a process called conjugation.
- Many prokaryotes are autotrophs; others are heterotrophs.

Viruses are not organisms; they cannot reproduce outside of cells. (pages 324--325)
- Every virus has the same basic structure: a core of DNA or RNA encased in a protein coat.
- Viruses are responsible for numerous lethal diseases of humans.

Eukaryotes have a nucleus and other interior compartments, which distinguishes them from prokaryotes. (pages 323-324)
- The endosymbiotic theory suggests mitochondria and chloroplasts may have been bacteria at one time.

Protists are eukaryotes that have a variety of forms, means of locomotion, mode of nutrition and method of reproduction. (pages 327-330)
- Fifteen protist phyla can be grouped into five categories based on common characteristics.

Fungi are very different from plants because they have filamentous bodies, cell walls of chitin, nuclear mitosis, and are heterotrophic. (pages 331-334)
- The method of reproduction is used to distinguish fungal phyla.
- Many fungi are important to terrestrial ecosystems because they act as decomposers.
- Some fungi form important symbiotic relationships with autotrophic organisms.

Key Terms Matching

1. _____ Miller-Urey experiment
2. _____ Bubble model
3. _____ Prokaryotes
4. _____ Bacteria
5. _____ Conjugation
6. _____ Binary fission
7. _____ Virus
8. _____ Endosymbiotic theory
9. _____ Protist
10. _____ Multicellularity
11. _____ Fungus
12. _____ Mycelium
13. _____ Spores
14. _____ Decomposers
15. _____ Mycorrhizae
16. _____ Lichen

a. Segments of DNA or RNA wrapped in a protein coat.
b. Plasmid exchange between bacteria.
c. Mutualism between a fungus and plant roots.
d. Symbiotic relationship between a fungus and algae or cyanobacteria.
e. One type of prokaryote.
f. Heterotrophic organism with a cell wall of chitin.
g. Explains the origin of mitochondria and chloroplasts.
h. Single-celled organisms that lack a nucleus.
i. Means by which prokaryotes reproduce.
j. Break down organic materials.
k. Mass of hyphae.
l. Suggests life evolved on the ocean's surface.
m. Highly variable group of eukaryotic organisms.
n. Common means of reproduction used by fungi.
o. Many cells with integration of activities.
p. Suggests life evolved in a primordial soup formed in the oceans.

17.1   How Cells Arose (Pages 318-320)

17. Life on earth originated approximately _____ years ago.
     a. 6,000   b. 60,000   c. 3.5 million   d. 2.5 billion   e. 10 billion

18. Tiny bubbles with an outer boundary similar to a cell membrane are called _____.

17.2   The Simplest Organisms (Pages 321-322)

19. Which of these structures facilitates conjugation?
    a. pilus   b. capsule   c. plasmid   d. flagellum   e. endospore

20. _____ made by some bacteria enable them to survive extreme environmental conditions.
    a. Pili   b. Capsules   c. Plasmids   d. Flagella   e. Endospores

21. Bacterial cell walls are characterized by
    a. silica.   b. peptidoglycan.   c. chitin.   d. cellulose.   e. They have no cell wall.

17.3   Comparing Prokaryotes to Eukaryotes (Page 323)

22. _____ Chemoautotrophs   a. Purple nonsulfur bacteria.
23. _____ Photoautotrophs   b. Includes the nitrifiers.
24. _____ Chemoheterotrophs   c. Most pathogens are in this group.
25. _____ Photoheterotrophs   d. Cyanobacteria.

17.4   Viruses Infect Organisms (Pages 324-325)

26. A major outbreak of what disease in 1918–19 killed 21 million people?
    a. rabies   b. influenza   c. AIDS   d. hepatitis B   e. ebola

27. Another name for this disease is viral encephalomyelitis.
    a. rabies   b. influenza   c. AIDS   d. hepatitis B   e. ebola

28. A disease that destroys immune defenses and results in death by infection or cancer is
    a. polio.   b. rabies.   c. AIDS.   d. ebola.

17.5   The Origin of Eukaryotic Cells (Page 326)

29. Mitochondria may have been _____ as suggested by the endosymbiotic theory.
30. Mitochondria still retain some of their own _____.

17.6   General Biology of Protists (Pages 327–328)

31. *Volvox* are an example of
    a. multicellularity.   d. a sexually reproducing organism.
    b. aggregation.   e. photosynthetic protozoans.
    c. a colonial organism.

32. Protists that ingest visible food particles are called _____ or _____.

33. Sexual reproduction in protists, although rare, can be carried out by
    a. gametic meiosis.
    b. zygotic meiosis.
    c. sporic meiosis.
    d. all of the above.

34. One key advantage to multicellularity is _____.

17.7   Kinds of Protists (Pages 329–330)

35. There are _____ distinct phyla of protists.

36. Nonmotile spore formers are _____.

37. The algae, dinoflagellates, euglenoids, and diatoms are all _____.

17.8   A Fungus Is Not a Plant (Pages 331-332)
True (T) or False (F) Questions
If you believe the statement to be false, rewrite the statement as a true one.

38. Fungi are autotrophs that are closely related to plants.
    Answer: _____ Restatement: _____
    _____

39. Molds consist of long chains of cells called mycelia and the walls dividing one cell from another are septa.
    Answer: _____ Restatement: _____
    _____

40. Fungal cell walls contain the complex polysaccharide, chitin.
    Answer: _____ Restatement: _____
    _____

17.9   Kinds of Fungi (Pages 333-334)

41. _____ Ascomycota   a. Many species may have lost the ability to reproduce sexually.
42. _____ Imperfect fungi   b. The greatest diversity of fungal species are classified in this phylum.
43. _____ Basidiomycota   c. Spores produced on club-shaped structures.
44. _____ Zygomycota   d. Hyphae lack septa.

45. In a lichen the fungal partner contributes the ability to _____ while the algae can _____ .
46. Fungi are probably the only organisms that can break down _____ .

True (T) or False (F) Questions
If you believe the statement to be false, rewrite the statement as a true one.

47. Mycorrhizae assist the growth of plants by absorbing nutrients from the soil.
    Answer: _____ Restatement: _____
    _____

Chapter Test

1. Oxygen began to accumulate in the early earth's atmosphere when
   a. methanogens began to colonize the ocean.   d. fungi evolved.
   b. cyanobacteria evolved.   e. viruses evolved.
   c. plants evolved.

2. A major difference between prokaryotic and eukaryotic cells is that
   a. most prokaryotes lack a cell wall, while most eukaryotes have one.
   b. all prokaryotes lack a nucleus, while most eukaryotes have one.
   c. all prokaryotes lack a nucleus, while all eukaryotes have one.
   d. only the prokaryotes can reproduce asexually.
   e. both a and d.

3. Endospores function to
   a. help the organism survive periods of drought.
   b. transmit genetic information from one bacterium to another.
   c. help the organism survive exposure to high temperatures.
   d. all of the above.
   e. both a and c.

4. At the end of conjugation, both of the cells involved
   a. form endospores.   d. have a copy of the plasmid DNA.
   b. perish.   e. are able to degrade most antibiotics.
   c. develop organelles.

5. A common cause of the common cold is the
   a. influenza virus.   b. polio virus.   c. rhinovirus.   d. HIV.   e. ebola virus.

6. Mitochondria probably evolved from
   a. ancestors of the phylum Chlorophyta.   d. aerobic, symbiotic eubacteria.
   b. aerobic forams.   e. archaebacteria.
   c. anaerobic, fermentative bacteria.

7. Endosymbiosis
   a. was only able to occur in the earth's early, oxygen-free atmosphere.
   b. probably led to all modern lines of eukaryotic organisms.
   c. has been disproved in several significant experiments.
   d. resulted in the division of the bacteria into two groups: the Eubacteria and the Archaebacteria.
   e. was unable to tolerate the pressures of natural selection.

8. The evolution of multicellularity
   a. was necessary to overcome surface-to-volume problems that larger cells would have encountered.
   b. was an advantage in avoiding prokaryotic predators.
   c. allowed organisms to more efficiently utilize oxygen.
   d. occurred shortly after the first cell formed.
   e. resulted in the first eulcaryotic organisms.

9. While walking through the woods in New England, you notice a crusty material on several of the rocks. You observe a slice of the material with a light microscope and notice long, filamentous strands of cells, many of which surround round, nucleated cells. What might this be?
   a. A mold   b. A colony of bacteria   c. A lichen   d. An amoeba   e. A mushroom

10. Fungi obtain their nutrients by
    a. photosynthesis.   d. external digestion.
    b. the ingestion of bacteria and protozoans. e. both a and b.
    c. internal digestion.

11. In lichens the fungal partner provides
    a. protection to the photosynthetic partner from the environment.
    b. carbohydrates to be used for nutrition.
    c. minerals and other nutrients.
    d. all of the above.
    e. both a and c.

Additional Study Help

Visit the ARIS (Assessment, Review, and Instruction System) site at aris.mhhe.com for quizzes, animations, and other study tools.

18 Evolution of Plants

Key Concepts Outline

Plants faced three key challenges in adapting to life on land: absorbing minerals, conserving water, and transferring gametes during sexual reproduction. (pages 338-339)
- Plants exhibit alternation of gametophyte (haploid) and sporophyte (diploid) generations.

The evolution of the plant kingdom can be traced by four key evolutionary innovations. (pages 340-341)
- These innovations are: alternation of generations, vascular tissue, seeds and flowers and fruits.

Gametophytes of nonvascular plants are the dominant generation and perform photosynthesis. (page 342)
- Nonvascular plants include: mosses, liverworts and hornworts; their sporophytes are not as conspicuous as the gametophytes and are usually nutritionally dependent on the gametophytes.

Vascular plants possess specialized water-conducting tissues. (page 343)
- Early vascular plants grew from their tips (primary growth), while later ones also grow in diameter (secondary growth).
- The life cycle of vascular plants is dominated by the sporophyte generation.

Ferns, lycopods, horsetails and whisk ferns are all seedless vascular plants. (pages 344-345)
- The gametophytes and sporophytes are both independent and self-sufficient.

Seeds greatly improve a plant's ability to successfully reproduce in a variety of environments. (pages 346-347)
- A seed is composed of a plant embryo, source of food and watertight covering.

Gymnosperms include the conifers, cycads, gnetophytes, and ginkgos. (pages 348-349)
- Gymnosperms produce seeds that are not enclosed in a fruit; gymnosperms do not produce flowers.

Angiosperms produce flowers that ensure pollen is transferred more accurately and seeds inside fruits that improve seed dispersal. (pages 350-354)
- Angiosperms, or "flowering plants," are the most successful of all plants.
- Different kinds of flowers are intended to attract different kinds of pollinators.
- Seeds in fruits are often dispersed when the fruit is eaten by an animal, but some fruits are dispersed by water or wind.

Key Terms Matching

1. _____ Cuticle
2. _____ Stomata
3. _____ Alternation of generations
4. _____ Sporophyte
5. _____ Gametophyte
6. _____ Moss
7. _____ Vascular tissue
8. _____ Primary growth
9. _____ Secondary growth
10. _____ Fern
11. _____ Seed
12. _____ Gymnosperm
13. _____ Cone
14. _____ Angiosperm
15. _____ Flower
16. _____ Endosperm
17. _____ Double fertilization
18. _____ Dicot

a. Cell division at the tips of stems and leaves.
b. Reproductive structure of angiosperms.
c. Cone producers; naked seed plants.
d. This allows plants to increase in diameter.
e. Produces two seed leaves.
f. Reproductive structure of pine trees.
g. This occurs in angiosperms.
h. The most successful of all plants.
i. Have soredia.
j. Triploid endosperm and embryo waiting for water.
k. Waxy outer covering that helps conserve water in plants.
l. Haploid, tiny stage in life cycle of most plants.
m. The gametophyte is the conspicuous stage in this life cycle.
n. Sporophyte produces gametophyte, gametophyte produces sporophyte.
o. The diploid state in a plant life cycle.
p. Nutrients for developing embryo.
q. Specialized openings in leaves that allow CO_2 to pass into the plant.
r. Conducts water, minerals and nutrients throughout plant tissues.

18.1   Adapting to Terrestrial Living (Pages 338-339)

19. A waxy covering called the _____ covering the surface of leaves and openings called _____ help plants conserve water.

20. The sporophyte generation is _____ (diploid or haploid) while the gametophyte generation is _____.

18.2   Plant Evolution (Pages 340-341)

21. The evolution of _____ allowed plants to grow larger.

22. Seeds function to
    a. transport water and nutrients throughout the plant.
    b. provide nutrients and protection for the embryo.
    c. aid in dispersal to more favorable environments.
    d. both b and c are correct

18.3   Nonvascular Plants (Page 342)
True (T) or False (F) Questions
If you believe the statement to be false, rewrite the statement as a true one.

23. The ferns were the first plants to evolve vascular tissue.
    Answer: _____ Restatement: _____
    _____

24. The common name for members of phylum Anthocerophyta is the liverworts.
    Answer: _____ Restatement: _____
    _____

18.4   The Evolution of Vascular Tissue (Page 343)

25. The earliest vascular plant for which we have a complete fossil is approximately _____ years old.
    a. 10 million   b. 100 million   c. 250 million   d. 300 million   e. 410 million

18.5   Seedless Vascular Plants (Pages 344–345)

26. A similarity between ferns and mosses is that both
    a. lack vascular tissue.   d. produce spores.
    b. reproduce with seeds.   e. both a and d
    c. have a dominant sporophyte stage.

27. In the fern life cycle the _____ and the _____ are produced on the gametophyte.
    a. archegonium, antheridium   d. sporophyte, seed
    b. sporophyte, vascular tissue   e. ovary, pollen
    c. sorus, seed

18.6   Evolution of Seed Plants (Pages 346–347)

28. In seed plants the male gametophyte is called the _____ or the _____.
29. A dormant, diploid plant embryo is known as the _____.

18.7 Gymnosperms (Pages 348–349)

30. An unusual feature of the ginkos is
    a. that the reproductive structures are produced on separate plants.
    b. the lack of flowers on the mature tree.
    c. their ability to decompose organic material.
    d. the lack of vascular tissue.
    e. their monocot seed.

18.8 Rise of the Angiosperms (Page 350)

31. Label the parts of the flower (a – g).

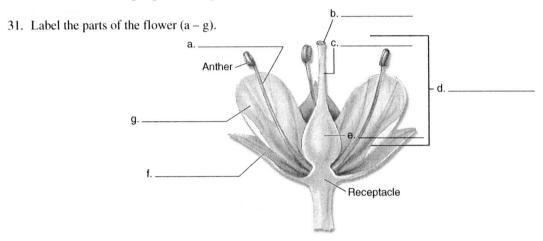

18.9 Why Are There Different Kinds of Flowers? (Page 351)

32. Flowers pollinated by bees are usually _____ (color) while red flowers are usually pollinated by _____.

33. Angiosperms, including oaks and birches, are pollinated by _____.

18.10 Improving Seeds: Double Fertilization (Pages 352–353)
True (T) or False (F) Questions
If you believe the statement to be false, rewrite the statement as a true one.

34. The endosperm is triploid and is a result of double fertilization.
    Answer: _____ Restatement: _____
    _____

35. Corn is a dicot plant.
    Answer: _____ Restatement: _____
    _____

18.11 Improving Seed Dispersal: Fruits (Page 354)

36. Three major vehicles of seed dispersement are _____ , _____ , and _____.

Chapter Test

1. Which of the following probably made the move to land by plants possible?
    a. the advent of multicellularity
    b. the accumulation of methane in the atmosphere
    c. the development of an ozone layer
    d. the development of the seed
    e. evolution of a flagellated sperm

2. Mycorrhizae
   a. enable a plant to photosynthesize more efficiently.
   b. are lichens.
   c. enable a plant to absorb many essential minerals.
   d. are respiratory structures found in some plants.
   e. both a and d.

3. Water conservation in plants is assisted by
   a.   the cuticle.   b. transpiration.   c. stomata.   d. cilia.   e. both a and c

4. The haploid stage in a plant life cycle is called the
   a. zygote.   b. embryo.   c. gametophyte.   d. sporophyte.   e. conspicuous stage.

5. A redwood tree represents which stage in the life cycle of that plant?
   a. zygote   b. embryo   c. gametophyte   d. sporophyte   e. none of the above

6. A difference between the gametophytes of nonvascular and vascular plants is that
   a. the nonvascular gametophyte is haploid, while the vascular gametophyte is diploid.
   b. the nonvascular gametophyte is bisexual, while the vascular plants have separate male and female gametophytes.
   c. the nonvascular gametophyte can only grow in water, while vascular gametophytes can grow on land.
   d. in nonvascular plants the gametophyte is the inconspicuous stage, while it is the conspicuous stage in
      vascular plants.
   e. there is no major difference between them.

7. One characteristic that modern mosses share in common with early plants is the
   a. need for water during fertilization.
   b. pollen grain.
   c. lack of a diploid stage.
   d. need for insects or wind for pollination.
   e. protected seed.

8. Why was vascular tissue such an important evolutionary advance?
   a. It made it easier for fertilization to occur on dry land.
   b. It allowed plants to have protected gametophytes.
   c. It was necessary for the formation of a seed coat.
   d. It allowed plants to grow larger and still conduct water to all of the tissue.
   e. Both a and d.

9. If the sporophyte was removed from the gametophyte in a Bryophyte, what would happen?
   a. The sporophyte would die because it is dependent on the gametophyte for nutrition.
   b. The gametophyte would die because it is dependent on the sporophyte for nutrition.
   c. The sporophyte would be able to survive on its own and it would produce another gametophyte.
   d. The sporophyte would form a protective capsule to help it survive until adequate nutrients were available.
   e. The sporophyte would continue to undergo meiosis.

10. Which of the following lack vascular tissue?
    a. mosses   b. the gymnosperms   c. ferns   d. angiosperms   e. the monocots

11. The cycads were the predominant land plants during the
    a. Cooksonian.   b. Jurassic.   c. Precambrian.   d. Cambrian.   e. none of the above.

12. The endosperm functions to
    a. protect the seed from desiccation.
    b. keep the seed dormant until water is available.
    c. provide nutrients to the young plant embryo.
    d. produce pollen grains.
    e. produce vascular tissue in the plant after the seed germinates.

Matching

13. _____ Frond   a. A plant embryo.
14. _____ Wood   b. A fern leaf.
15. _____ Seed   c. Fern sporophytes release these.
16. _____ Spore   d. This is the result of secondary plant growth.

Additional Study Help

Visit the ARIS (Assessment, Review, and Instruction System) site at aris.mhhe.com for quizzes, animations, and other study tools.

19 Evolution of Animals

Key Concepts Outline

Animals are complex multicellular heterotrophs that are mobile and sexually reproduce. (pages 358-359)
- Animal cells lack cell walls and their forms and habitats are diverse.

There are five key body transitions that characterize animal evolution and distinguish the various animal phyla. (pages 360-363)
- The five transitions are evolution of: tissues, bilateral symmetry, a body cavity, deuterostome development, and segmentation.

Sponges and Cnidarians are the simplest animals. (pages 364-365)
- Sponges lack specialized tissues and symmetry; cnidarians have specialized tissues and radial symmetry.

Cephalization is usually associated with bilateral symmetry; cephalized animals are mobile and able to sense the presence of food, danger, or mates as they move through their environment head first. (pages 366-367)
- Flatworms are some of the simplest animals with bilateral symmetry; they have internal organs and a head, but lack a body cavity.

Development of a body cavity was a key transition in the evolution of animals. (pages 368-373)
- A pseudocoelom is a body cavity between the endoderm and the mesoderm; a coelom is a body cavity entirely within the mesoderm.
- Roundworms have a pseudocoelom and lack segmentation.
- Mollusks have a coelom and lack segmentation; annelids and arthropods have coeloms and are segmented.

Animals classified as echinoderms and chordates are deuterostomes; previously covered coelomates are protostomes. (pages 374-376)
- Adult echinoderms have radial symmetry and echinoderms have an endoskeleton.
- Chordates are characterized by a notochord which becomes a backbone in adult vertebrates.

Animal phyla diversification occurred in the sea; jawless fish were the first vertebrates to evolve. (page 377)
- Animals that successfully invaded the land include: arthropods, amphibians, reptiles, birds and mammals.
- Periodic mass extinctions have occurred; the greatest, at the end of the Paleozoic era, rendered 90% of all species extinct.

Fishes are characterized by gills that enable them to extract oxygen from the water, and a single-loop circulatory system (pages 378-379).
- Bony fish are quite successful and are characterized by swim bladders and lateral line systems.

Amphibians and reptiles successfully invaded land thanks to legs, lungs and pulmonary and systemic circulation. (pages 380-381)
- Reptiles are well suited for life on land thanks to the production of an amniotic egg, dry skin and thoracic breathing.

Characteristics of birds such as feathers and a strong, but light skeleton enable them to fly. (page 382)
- The fossil record includes *Archaeopteryx*, the oldest bird, was similar to small dinosaurs.

Mammals are endotherms characterized by mammary glands and hair. (pages 383-384)
- There are different forms of embryonic development in mammals.

Key Terms Matching

1. _____ Phylum
2. _____ Choanocytes
3. _____ Radial symmetry
4. _____ Nematocyst
5. _____ Acoelomates
6. _____ Pseudocoelomates
7. _____ Radula
8. _____ Segmentation
9. _____ Exoskeleton
10. _____ Protostome
11. _____ Deuterostome
12. _____ Notochord
13. _____ Paleozoic era
14. _____ Mass extinctions
15. _____ Gill
16. _____ Placental mammal
17. _____ Swim bladder
18. _____ Amniotic egg
19. _____ *Archaeopteryx*
20. _____ Monotreme
21. _____ Marsupial

a. Lacking a body cavity.
b. Characteristic of chordates.
c. Developmental fate of each cell is fixed when it appears.
d. Specialized cells in sponges.
e. Tongue of mollusks.
f. Made of chitin.
g. The symmetry of primitive eumetazoans.
h. The echinoderms were the first of these.
i. Cnidocyte harpoon.
j. A solid body cavity.
k. A group larger than class.
l. The annelid worms were the first to try this.
m. Extract dissolved oxygen from water.
n. Duckbill platypus.
o. Almost all animals alive today originated in the sea at this time.
p. Transferred to a pouch after birth.
q. Four of these occurred in the Paleozoic era.
r. A true flying dinosaur.
s. A gas-filled sac in bony fishes.
t. They spend much time nurturing their young.
u. Contains a bird embryo.

19.1 General Features of Animals (Pages 358–359)

22. _____ Gastrula
23. _____ Morula
24. _____ Blastula
25. _____ Blastopore

a. It's from this stage that the digestive tract develops.
b. A hollow ball of cells.
c. An opening in the embryo.
d. A solid ball of cells.

19.2 Five Key Transitions in Body Plan (Pages 360-363)

26. Phylogenies have traditionally been based upon anatomical features and embryological development. Today the field of _____ _____, using RNA and DNA sequencing data, may significantly alter relationships among the animal phyla.

19.3 Sponges and Cnidarians: The Simplest Animals (Pages 364–366)

27. Although subkingdom Parazoa is classified in kingdom Animalia, its members differ from the Eumetazoans in the following way(s):
    a. The parazoans have unique organ systems not seen in the eumetazoans.
    b. The parazoans lack tissues and organs.
    c. The parazoans lack a definite symmetry.
    d. The parazoans only reproduce asexually.
    e. both b and c.

28. Protists called _____ were probably the ancestors of sponges and may have been the ancestors of all animals.
    a. nematocysts   b. cnidarians   c. cnidocytes   d. choanoflagellates   e. ectoderms

29. Cnidarians are carnivorous and capture their prey with stinging cells called _____, each of which contain a(n) _____.

30. Sponges lack _____ which is present in the eumetazoans.

19.4   The Advent of Bilateral Symmetry (Pages 366-367)

31. Parasitic flatworms are protected from the digestive enzymes of their host by
    a. a hard cyst.   d. enzymes of their own.
    b. endospores.   e. none of the above
    c. resistant epithelial layers.

32. The flatworms are the simplest organisms in which _____ occur.
    a. discrete reproductive structures   d. organs
    b. sensory structures   e. eggs
    c. a nervous system

33. Flame cells are
    a. excretory structures seen in the flatworms.   d. used to catch prey.
    b. reproductive organs characteristic of eumetazoans.   e. part of the flatworm circulatory system.
    c. required for cell division in the flatworms.

19.5   The Advent of a Body Cavity (Pages 368–373)

34. In pseudocoelomates a body cavity develops _____ while in the coelomates the body cavity develops _____.

35. _____ Whiplike bodies   a. Rotifers
36. _____ Common in the soil   b. Nematodes
37. _____ They have a crown of cilia on their head
38. _____ Stylets

39. The mollusks are the only major phylum of coelomates that lack _____.
40. Specialized tissues develop by _____ in the animals.
41. A great advantage of segmentation is the _____ it offers.

42. Arthropods have not been able to achieve great size because
    a. the type of organ systems they possess could not support such a development.
    b. they would be unable to successfully reproduce.
    c. chitin is brittle and must be thick to bear the pull of muscles.
    d. their vision is not good enough to hunt larger prey.
    e. none of the above

43. Of the chelicerates, which of the following transmits Lyme disease to humans?
    a. spiders   b. mites   c. scorpions   d. ticks   e. annelids

19.6   Redesigning the Embryo (Pages 374–377)

44. During embryonic development, the progressive division of cells is called _____.
45. In deuterostomes the egg cleaves _____ and in protostomes the egg cleaves _____.

True (T) or False (F) Questions
If you believe the statement to be false, rewrite the statement as a true one.

46. Echinoderms use a hydraulic system called the water vascular system to move.
    Answer: _____ Restatement: _____
    _____

47. Echinoderms are bilaterally symmetrical as larvae but become radially symmetrical as adults.
    Answer: _____ Restatement: _____
    _____

48. All chordates exhibit four characteristics at some point in their life cycle. They are
    _____, _____, _____ and _____.

19.7   Overview of Vertebrate Evolution (Page 377)

49. The first organisms to invade land were the _____.
50. Diversification of animal phyla occurred in the _____.

19.8   Fishes Dominate the Sea (Pages 378-379)

51. Four characteristics of fishes are _____, _____, _____, _____.
52. The swim bladder of fishes evolved as a(n) _____.

19.9   Amphibians and Reptiles Invade the Land (Pages 380-381)

53. Amphibians respire by using lungs and
    a. gills.   b. their skin.   c. tracheids.   d. sieve tubes.   e. both a and c.

54. Development of a(n) _____ allowed amphibians to more efficiently deliver oxygen to their muscles.
    a. pulmonary vein   d. smaller body size
    b. closed circulatory system   e. true lung
    c. four-chambered heart

19.10   Birds Master the Air (Page 382)

55. _____ Kiwis   a. Only two toes.
56. _____ Ducks   b. Well developed vocal organs.
57. _____ Parrots   c. Broad bill with filtering ridges.
58. _____ Ostriches   d. Live only in New Zealand.

19.11   Mammals Adapt to Colder Times (Pages 383-384)

59. _____ Anteaters   a. Order Perissodactyla.
60. _____ Lemurs   b. Order Cetacea.
61. _____ Horses   b. Opposable thumb.
62. _____ Dolphins   d. Many lack teeth.

Chapter Test

1. Animal(s)
   a. cells lack cell walls.   d. are heterotrophic or autotrophic.
   b. sometimes have cell walls, such as in the arthropods.   e. are found primarily on land.
   c. are always mobile.

2. Choanocytes are
   a. primitive organs in sponges that function in digestion.
   b. flagellated cells that draw water through the body cavity of sponges.
   c. amoeba-like cells that wander over the surface of sponges and distribute nutrients to other cells.
   d. calcium deposits that provide structure to sponges.
   e. both a and b.

3. One property that sponges share in common with many other animals is
   a. a life cycle that involves alternation of generations.   d. the ability to digest cellulose.
   b. a free-swimming medusa in its life cycle.   e. none of the above
   c. cell recognition.

4. Cnidarians project a nematocyst to capture their prey by
   a. ejecting it with a jet of water.
   b. using a springlike apparatus.
   c. building up a high internal osmotic pressure.
   d. employing simple muscle fibers.
   e. coiling and releasing the tendrils on which the nematocysts are found.

5. The cnidarians have which of the following evolutionary advances over the sponges?
   a. bilateral symmetry
   b. cephalization
   c. a life cycle involving a dominant haploid form
   d. extracellular digestion
   e. a nonmotile mature form

6. Which of the following is an example of an organism with the medusae body form?
   a. a sponge   b. a hydra   c. a coral   d. an amoeba   e. a jellyfish

7. Most cnidarians
   a. exist only as medusae in their life cycle.
   b. exist only as polyps in their life cycle.
   c. encyst when they encounter adverse environmental conditions.
   d. alternate between a polyp and a medusa during their life cycle.
   e. none of the above.

8. Cephalization is
   a. an evolutionary trend in the cnidarians.
   b. the evolution of a head.
   c. the lack of a distinctive head seen in the primitive animal phyla.
   d. the trend towards bilateral symmetry seen in the chordates.
   e. missing in many of the animal phyla.

9. An example of an animal with a pseudocoelomate body is (are)
   a. hydra.   b. planaria.   c. rotifers.   d. Dugesia.   e. jellyfish.

10. A radula is
    a. a sharp structure that is injected into the prey of a mollusk.
    b. a protective coating made of calcium carbonate on sponges.
    c. a small, internal shell found in cephalopods.
    d. a rasping, tonguelike organ in mollusks.
    e. necessary for mollusks to be motile.

11. The segments of annelids are
    a. actually one large unit internally.   d. not present in all species.
    b. partitioned internally.   e. both b and c.
    c. specialized for different functions.

12. Rocky Mountain spotted fever and Lyme disease are transmitted by a(n)
    a. insect.   b. protozoan.   c. fungus.   d. nematode.   e. arachnid.

13. How do the crustaceans differ from the insects?
    a. Crustaceans have an exoskeleton made of chitin.
    b. Crustaceans have legs on their abdomen and thorax.
    c. Crustaceans have jointed appendages.
    d. Crustaceans are exclusively found in marine habitats.
    e. Both a and c.

14. Malpighian tubules
    a. assist echinoderms in motility.
    b. regulate osmotic pressure in insect cells.
    c. play a major role in digestion in insects.
    d. function in excretion in insects.
    e. secrete the exoskeleton in insects.

15. Animals in which the blastopore becomes the mouth are called
    a. stomates.   b. deuterostomes.   c. protostomes.   d. echinostomes.   e. prestomes.

16. The significance of a notochord in the evolution of chordates is that it
    a. provided an internal attachment point for muscles.
    b. allowed for the development of a more complex nervous system.
    c. eliminated the need for segmentation.
    d. allowed the organism to grow larger.
    e. both a and c.

17. Which of the following might be a reason that the modern fishes were more successful than the agnathans?
    a. The agnathans were only able to feed on foods found on the bottom of the ocean, while modern fishes can take advantage of more diverse food sources.
    b. Modern fishes were hunters and sought food in diverse places, while the agnathans were limited to filter feeding on the ocean's bottom.
    c. Modern fishes were more efficient at removing oxygen from the water.
    d. The agnathans were unable to produce the large number of offspring produced by modern fishes.
    e. None of the above.

18. A bony skeleton versus bony body plates in fish is an advantage because it
    a. supported strong muscles that enabled fish to chase prey.
    b. provided better protection from predators.
    c. allowed fish to grow larger and become more efficient predators.
    d. allowed fish to swim in the upper levels of the water.
    e. allowed fish to survive in freshwater habitats.

19. A swim bladder allows a fish to
    a. regulate its water content in seawater.
    b. regulate its salt content in seawater.
    c. regulate its salt content in freshwater.
    d. control its buoyancy in water.
    e. tolerate changes in water pressure.

20. The monotremes are classified as mammals because they
    a. are endoderms.
    b. are carnivores.
    c. are omnivores.
    d. have mammary glands.
    e. give birth to live young rather than lay eggs.

21. Fish resist dehydration in a marine environment by
    a. maintaining a higher internal concentration of salt than is in the water.
    b. reabsorbing water through their kidneys.
    c. secreting a slime layer to cover their body.
    d. both a and c.
    e. none of the above.

Additional Study Help

Visit the ARIS (Assessment, Review, and Instruction System) site at aris.mhhe.com for quizzes, animations, and other study tools.

20   Ecosystems

Key Concepts Outline

Ecology is the study of how organisms fit into and interact with their environment. (pages 388-392)
- An ecosystem is a self-sustaining group of organisms and the minerals, water, and weather that make up their habitat.
- Energy flows through an ecosystem from producers to consumers to detritivores and decomposers in a food web; organisms are classified at a trophic or feeding level based on how they get their energy.
- The biomass of producers tends to be highest in an ecosystem because of the loss of energy as heat at each step of a food chain.

The health of ecosystems is dependent upon the proper cycling of four materials. (pages 393-397)
- Water cycles either by evaporation and condensation or by absorption and transpiration.
- Carbon begins cycling through living things when it is captured during photosynthesis; it continues cycling via respiration, combustion and erosion.
- Nitrogen cycling is dependent upon the ability of some bacteria to fix nitrogen; phosphorous is available in the soil and dissolved in water and can be the limiting factor that determines what organisms are able to live in a certain ecosystem.

The climate of a particular area is affected by the intensity of the sun's rays, air currents, elevation, and ocean currents. (pages 398-399)
- A rain shadow effect is caused by mountains, which force winds upwards, causing them to cool and release their moisture on the windward side of mountains.
- World climate can be profoundly affected by disturbances in ocean currents.

Earth is comprised of a variety of aquatic and terrestrial ecosystems. (pages 402-410)
- There are three principal ocean ecosystems: intertidal shallows, open-sea surface and deep-sea waters.
- Freshwater ecosystems cover only a very small percentage of the earth's surface and are closely tied to terrestrial ecosystems.
- There are a number of different terrestrial ecosystems, also called biomes; the communities of these biomes are largely determined by patterns of rainfall and temperature.

Key Terms Matching

1. _____ Ecology
2. _____ Habitat
3. _____ Ecosystem
4. _____ Producers
5. _____ Consumers
6. _____ Food web
7. _____ Evaporation
8. _____ Transpiration
9. _____ Fossil fuel
10. _____ Nitrogen fixation
11. _____ Rain shadow
12. _____ Plankton
13. _____ Biome

a. A complicated path of energy flow.
b. Photosynthetic organisms.
c. Coal and oil.
d. Evaporation from leaf surfaces.
e. Community + habitat.
f. A terrestrial ecosystem.
g. Tiny organisms many fish feed on.
h. An effect that is responsible for Death Valley.
i. How organisms interact with each other and their physical environment.
j. Water entering the atmosphere when heated by the sun.
k. Everyone but autotrophs.
l. Binding nitrogen to hydrogen.
m. Where a community lives.

20.1   Energy Flows Through Ecosystems (Pages 388–391)

14. The vast majority of energy taken into an ecosystem is
    a. converted into biomass by plants.
    b. utilized by secondary consumers.
    c. lost as heat.
    d. used by the primary consumers.
    e. concentrated in the decomposers.

15. Although the biomass of a rain forest is much greater than the biomass of a cornfield
    a. the number of primary consumers in the cornfield exceeds those in the rain forest.
    b. the rain forest's net primary productivity is much lower in relation to its biomass.
    c. the cornfield is a more stable ecosystem.
    d. the cornfield cannot survive without nutritional supplementation.
    e. none of the above are true.

16. Which of the following is a difference between a food chain and a food web?
    a. Food chains involve only plants while food webs involve animals.
    b. Food chains involve only plants while food webs involve both plants and animals.
    c. Food chains involve plants and animals but food webs involve only animals.
    d. Food chains are linear and food webs are complex.
    e. Food chains include decomposers but food webs do not.

17. The carnivores are classified as secondary consumers because they eat
    a. more than one species of herbivore.
    b. herbivores or other carnivores.
    c. herbivores.
    d. so often.
    e. both a and d.

20.2   Ecological Pyramids (Page 392)

18. Inverted biomass pyramids are most likely to occur in_____.
19. In the food chain the biomass of the _____ tends to be greater than any other level.

20.3   The Water Cycle (Pages 393–394)

20. Most water lost from plants is by means of _____.
21. Water cycles through ecosystems by _____ and _____.

20.4   The Carbon Cycle (Page 395)

True (T) or False (F) Questions
If you believe the statement to be false, rewrite the statement as a true one.

22. Carbon captured from the atmosphere by photosynthesis can only return through respiration.
    Answer: _____ Restatement: _____
    _____

23. Plants represent the most plentiful source of carbon in the ecosystem.
    Answer: _____ Restatement: _____
    _____

20.5 Soil Nutrients and Other Chemical Cycles (Pages 396–397)

24. Animals obtain nitrogen by ingesting _____.

25. Plant growth is most likely to be limited by scarcity of _____ in the environment.
    a. sulfur   b. nitrogen   c. phosphorus   d. carbon   e. both b and c

20.6 The Sun and Atmospheric Circulation (Page 398)

True (T) or False (F) Questions
If you believe the statement to be false, rewrite the statement as a true one.

26. The sun drives circulation of the atmosphere.
    Answer: _____ Restatement: _____
    _____

27. The major atmospheric circulation patterns result from the interactions between six large air masses.
    Answer: _____ Restatement: _____
    _____

20.7 Latitude and Elevation (Page 399)

28. A rain shadow is likely to develop
    a. in almost any very hot climate.
    b. at elevations exceeding 4,500 ft.
    c. on the leeward side of a mountain.
    d. on the north slope of a mountain.
    e. near the equator.

20.8 Patterns of Circulation in the Ocean (Pages 400–401)

29. _____ are huge spiral patterns of circulation that occur in oceans.
30. The cause of changes in the world's weather currents such as _____ are _____.

20.9 Ocean Ecosystems (Pages 402–403)

31. The greatest diversity of species in the oceans occurs in_____.
32. Rapid turnover of nutrients takes place in the oceans and is due to _____.

20.10 Freshwater Ecosystems (Pages 404–405)

33. _____ Scarce nutrients   a. Eutrophic lakes
34. _____ Abundant organic matter   b. Oligotrophic lakes
35. _____ Little oxygen in summer
36. _____ Deep waters rich in oxygen

20.11 Land Ecosystems (Pages 406–410)

37. The world's biomes are broken up by _____, _____ and _____.

Chapter Test

1. Ecology is the study of
   a. how organisms interact with their environment.
   b. where we live.
   c. how organisms interact with each other.
   d. the different environments in the world.
   e. all of the above.

2. A variety of different bacteria and yeasts live on human skin. Together they are considered to be a(n)
   a. ecosystem.   b. population.   c. community.   d. trophic level.   e. biome.

3. Photosynthetic bacteria are in trophic level
   a. 1.   b. 2.   c. 3.   d. all of the above.   e. none of the above.

4. Molds that break down leaves and other dead organic material on the forest floor are in trophic level
   a. 1.   b. 2.   c. 3.   d. all of the above.   e. none of the above.

5. A lion that feeds on zebras is in trophic level
   a. 1.   b. 2.   c. 3   d. all of the above.   e. none of the above.

6. Herbivores are _____ of ecosystems.
   a. primary consumers.   b. decomposers.   c. producers   d. secondary consumers   e. tertiary consumers

7. Primary productivity refers to the
   a. rate of photosynthesis in plants and photosynthetic bacteria as compared to the rate of plant consumption by herbivores.
   b. total amount of light energy converted to organic compounds in a given area per unit time.
   c. rate of decomposition by the detritivores.
   d. total biomass of the photosynthetic organisms in the ecosystem.
   e. none of the above.

8. In the earth's ecosystems, energy
   a. is recycled and is never really lost.
   b. flows from producers to consumers and back to producers.
   c. flows in one direction, from producers to consumers.
   d. is produced at all trophic levels.
   e. both a and d.

9. A consequence of cutting down forests is
   a. that water is no longer returned to the atmosphere over the area of the forest.
   b. the loss of animal habitats.
   c. the production of high-quality agricultural land.
   d. all of the above.
   e. both a and b.

10. Burning wood
    a. releases oxygen into the atmosphere.   d. releases carbon into the atmosphere.
    b. consumes carbon dioxide.   e. releases nitrogen into the atmosphere.
    c. destroys the carbon that was in the wood.

11. Nitrogen fixation is accomplished by
    a. bacteria in the atmosphere.
    b. bacteria in the soil and in plant root nodules.
    c. plants.
    d. the fungi.
    e. all of the above.

12. Why did their corn grow better when Native Americans added a fish to the soil at planting time?
    a. The fish added carbon to the soil.
    b. The fish added "fixed" nitrogen to the soil.
    c. The fish added phosphorus and sulfur to the soil.
    d. The fish provided the vitamins that the young plant required.
    e. all of the above.

13. Pollution of lakes by commercial detergents that contain phosphates
    a. kills the bacteria there and causes a breakdown in the food chain.
    b. encourages the growth of bacteria and an increase in the fish population.
    c. encourages the growth of algae, which leads to the suffocation of fish and other animals.
    d. raises the pH.
    e. lowers the pH.

14. Ocean currents are determined by
    a. proximity to land.
    b. underwater geography.
    c. atmospheric circulation.
    d. the season.
    e. none of the above.

15. Deep sea waters below 300 meters
    a. harbor a rich variety of life, including the red algae.
    b. are rich in photoplankton.
    c. consist mostly of plant life.
    d. harbor limited kinds of life, some of which are very strange.
    e. both b and c.

16. Oligotrophic lakes are
    a. only found at high latitudes.
    b. saltwater lakes.
    c. rich in organic matter and nutrients.
    d. scarce in organic matter and nutrients.
    e. rich in plant life.

Matching

17. _____ Tundra
18. _____ Tropical rain forest
19. _____ Grasslands
20. _____ Chaparral
21. _____ Savannas
22. _____ Deciduous forests
23. _____ Desert

a. The richest ecosystems on earth are found here.
b. Many of the animal residents live in burrows.
c. Dry climates that border tropics.
d. Low-growing, spiny plants; Mediterranean climate.
e. Deer, rabbits, raccoons, and squirrels.
f. Permafrost may be a meter deep.
g. They are also called prairies.

Additional Study Help

Visit the ARIS (Assessment, Review, and Instruction System) site at aris.mhhe.com for quizzes, animations, and other study tools.

21   Populations and Communities

Key Concepts Outline

Many populations exhibit a sigmoid growth curve, with a relatively slow start in growth, a rapid increase, and then a leveling off when carrying capacity of the environment is reached. (pages 414-418)
- Population growth is regulated by density-independent and density-dependent effects.
- Populations with r-selected adaptations reproduce early and have many offspring; populations with K-selected adaptations tend to have fewer offspring and slower growth rates.
- Survivorship curves express the age distribution characteristics of populations and the tendency for mortality to be concentrated among the young, among the old or to be independent of age.

Communities are the various species that occur in a certain area; ecological and evolutionary processes are affected by the interactions among the species. (pages 420-433)
- Each species plays a specific role called a niche in its ecosystem; resource partitioning reduces competition between similar species, allowing them to coexist.
- Coevolution resulted in various forms of symbiosis: mutualism; commensalism; and parasitism, and a variety of predator-prey interactions such as defensive coloration, chemical defenses and mimicry.

Ecological succession occurs when one community is replaced by another more complex community. (pages 434-435)
- Primary succession takes place in barren areas.
- Secondary succession takes place in areas where the original communities of organisms have been disturbed, often by human activities.

Key Terms Matching

1. _____ Population
2. _____ Carrying capacity
3. _____ Demography
4. _____ Niche
5. _____ Competition
6. _____ Coevolution
7. _____ Symbiotic relationship
8. _____ Commensalism
9. _____ Mutualism
10. _____ Parasitism
11. _____ Aposematic coloration
12. _____ Cryptic coloration
13. _____ Batesian mimicry
14. _____ Succession
15. _____ Species richness
16. _____ Biodiversity

a. When two organisms try to use the same resource.
b. Coloration used to warn off predators.
c. For example, symbiosis.
d. The statistical study of populations.
e. Living together.
f. Helps an organism blend into its environment.
g. One community replacing another over time.
h. One participant benefits at the expense of the other.
i. How the number of species in an ecosystem are measured.
j. Where a population eventually stabilizes.
k. Two organisms living together, both benefit from the relationship.
l. The number of species in an ecosystem.
m. One participant benefits, the other neither benefits nor is harmed.
n. Looking like someone who tastes bad.
o. Groups of individuals of the same species living together.
p. All of the ways an organism utilizes the resources in its environment.

21.1   Population Growth (Pages 414–415)

17. Population growth levels off due to
    a. an increase in the average age of population members and a resultant lessening in reproduction.
    b. competition for shelter, food, light or other factors.
    c. an increase in emigration from the population.
    d. higher incidence of lethal mutations.
    e. both a and c.

18. Most natural populations exhibit _____ growth.
    a. exponential   b. varied   c. seasonal   d. logistic   e. checked

21.2   The Influence of Population Density (Page 416)

19. _____ effects such as weather regulate the growth of populations regardless of _____.
20. Salmon harvesters try to harvest fish at the population's _____ so as not to damage the productivity of the population.
21. When population size increases to a critical point, _____ effects become more important in the regulation of the population's size.

21.3   Life History Adaptations (Page 417)

22. A characteristic of K selected populations is
    a. long lifespan and the production of a few offspring.
    b. reproduction at an early age and little or no parental care.
    c. a long maturation time and a high mortality rate.
    d. reproducing (usually) only once and slow maturation.
    e. a high mortality rate.

21.4   Population Demography (Page 418)

23. _____ Survivorship curve   a. A group of individuals of the same age.
24. _____ Cohort   b. The number of offspring in a standard time.
25. _____ Sex ratio   c. Percentage of an original population to survive to a given age.
26. _____ Fecundity   d. Often influenced by number of females in a population.

21.5   Communities (Page 420)

True (T) or False (F) Questions
If you believe the statement to be false, rewrite the statement as a true one.

27. Communities represent groups of species that have interacted and evolved together over long periods of time.
    Answer: _____ Restatement: _____
    _____

28. Communities exist in a particular place because the ranges of its species overlap in that place.
    Answer: _____ Restatement: _____
    _____

21.6   The Niche and Competition (Pages 421–423)

29. A fundamental niche is
    a. defined as the way in which an organism uses its environment.
    b. the niche an organism is theoretically capable of using.
    c. the niche an organism uses in the presence of competitors.
    d. the basic and minimal requirements for survival of a species.
    e. none of the above.

30. A niche is described as
    a. the space an organism occupies.
    b. the food an organism consumes.
    c. an organism's moisture requirements.
    d. the temperature at which an organism can survive.
    e. all of the above.

True (T) or False (F) Questions
If you believe the statement to be false, rewrite the statement as a true one.

31. Gause's principle states that two species can occupy the same area indefinitely.
    Answer: _____ Restatement: _____

    _____

32. Extinction is the inevitable conclusion of two species indefinitely occupying the same niche.
    Answer: _____ Restatement: _____

    _____

33. _____ species are more likely to exhibit significant morphological differences.
34. Species occupying the same geographic area are said to be _____.

21.7 Coevolution and Symbiosis (Pages 424–427)

True (T) or False (F) Questions
If you believe the statement to be false, rewrite the statement as a true one.

35. Symbiosis, or two species living together, is seldom seen in nature.
    Answer: _____ Restatement: _____

    _____

36. In commensalism one species benefits at the expense of the other species.
    Answer: _____ Restatement: _____

    _____

37. Which of the following is an example of commensalism?
    a. A tapeworm living in the gut of its host.
    b. A bird feeding on insects on the back of a large herbivore.
    c. A lichen.
    d. Ants living on a plant.
    e. All of the above are examples of commensalism.

38. Ants live on the *Acacia* plant's _____, which they use as their primary source of food.
    a. honeydew   b. nectar   c. cellulose   d. Beltian bodies   e. starch

39. _____ are insects that lay their eggs on a living host which is usually killed when they hatch.
40. The laying of eggs in the nest of another species of birds is known as _____ parasitism.

21.8 Predator–Prey Interactions (pages 428–429)

True (T) or False (F) Question
If you believe the statement to be false, rewrite the statement as a true one.

41. Keystone species are important because their presence affects the composition and function of their communities.
    Answer: _____ Restatement: _____

    _____

42. Predation reduces competition in a community by
    a. reducing the number of species living there.
    b. stimulating the production of more offspring to ensure the survival of the prey species.
    c. reducing the number of individuals competing for resources.
    d. interacting with several species in the community.
    e. all of the above.

43. As the number of individuals of the prey species increases, the predator population
    a. decreases.   b. increases.   c. stabilizes.   d. evolves.   e. none of the above.

21.9   Plant and Animal Defenses (Pages 430–431)

44. Plants that have evolved secondary compounds to protect themselves from herbivores
    a. will not be consumed by any species of organism and will have a competitive advantage over other plants.
    b. are often the exclusive food for some species that has adapted to break down the secondary compound.
    c. are rare in nature.
    d. both a and b are true.
    e. both b and c are true.

45. Some animal species have evolved to defend themselves from predators with _____, _____
    or _____.

21.10   Mimicry (Pages 432–433)

46. In order for mimicry to be effective in protecting a species from predation, it must be
    a. rare relative to the model species.
    b. common in several insect species in the area.
    c. based on the production of an unpleasant-tasting substance.
    d. based on the production of a toxic substance.
    e. counteracted by the production of sufficient offspring to offset those that are consumed.

21.11   Ecological Succession (Pages 434–435)

47. _____ Tolerance   a. A species is able to survive in less than favorable conditions.
48. _____ Facilitation   b. Changes in the habitat create unfavorable conditions for another species.
49. _____ Inhibition   c. One species creates changes in the habitat that favor another species.

Chapter Test

1. Population dispersion may be
   a. uniform.   b. clumped.   c. random.   d. all of the above.   e. a and b only.

2. The rapid growth of bacteria that occurs after inoculation into agar in a petri plate is called a(n)
   a. bloom.   d. accelerated growth.
   b. lawn.   e. activated growth.
   c. exponential growth.

3. A rise in the incidence of tuberculosis in crowded prisons is an example of a(n)
   a. survivorship effect.   d. limiting factor.
   b. density-dependent effect.   e. density-independent effect.
   c. dispersion effect.

4. The age distribution of a population influences
   a. the carrying capacity.   d. dispersion.
   b. survivorship curve.   e. emigration of individuals.
   c. intrinsic rate of increase.

5. In order to protect themselves from herbivores, plants may
   a. contain toxic substances.   d. have spines.
   b. contain unpleasant-tasting substances.   e. all of the above.
   c. have thorns.

6. Lichens are an example of
   a. parasitism.   b. commensalism.   c. mutualism.   d. synergism.   e. none of the above.

7. Bees that pollinate flowers are an example of
   a. parasitism.   b. commensalism.   c. mutualism.   d. synergism.   e. none of the above.

8. Aphids living on roses are an example of
   a. parasitism.   b. commensalism.   c. mutualism.   d. synergism.   e. none of the above.

9. Brightly colored poison dart frogs are an example of
   a. cryptic coloration.   d. Müllerian mimicry.
   b. aposematic coloration.   e. none of the above.
   c. Batesian mimicry.

10. The monarch butterfly is an example of
    a. cryptic coloration.   d. Müllerian mimicry.
    b. aposematic coloration.   e. none of the above.
    c. Batesian mimicry.

11. Tomato worms are exactly the same color as the plants that they feed on. This is an example of
    a. cryptic coloration.   d. Müllerian mimicry.
    b. aposematic coloration.   e. none of the above.
    c. Batesian mimicry.

12. Lichens growing on the surface of rocks provide an example of
    a. Batesian mimicry.   d. a climax community.
    b. primary succession.   e. a keystone species.
    c. secondary succession.

13. The growth that occurs after a forest fire has occurred is an example of
    a. regrowth.   d. a climax community.
    b. primary succession.   e. keystone speciation.
    c. secondary succession.

Additional Study Help

Visit the ARIS (Assessment, Review, and Instruction System) site at aris.mhhe.com for quizzes, animations, and other study tools.

22   Behavior and the Environment

Key Concepts Outline

The way an animal responds to stimuli in its environment is animal behavior. (pages 440-442)
- Ethologists study animal behavior in natural settings and try to determine the physiological mechanisms that produce behavior.
- The investigation of how genes influence behavior is called behavioral genetics; the key role genes play in causing many behaviors is supported by studies of many animals, including humans.

Learning is the process that occurs when an animal alters its behavior because of previous experience. (pages 443-445)
- Examples of nonassociative learning are sensitization and habituation.
- Conditioning and imprinting are examples of associative learning.
- Some research indicates animals can reason.

The study of how natural selection affects behavior is called behavioral ecology. (pages 446–456)
- Natural selection favors foraging and territorial behaviors that enable an animal to acquire more energy; successful migratory behavior; characteristics that maximize reproductive success; successful communication and altruistic behavior.

Human behavior is influenced by genes, learning and experience.

Key Terms Matching

1. _____ Ethology
2. _____ *fosB* alleles
3. _____ Operant conditioning
4. _____ Cognitive behavior
5. _____ Territoriality
6. _____ Migration
7. _____ Dance language
8. _____ Altruism

a. Behavior associated with reward or punishment.
b. Movement over long distances to breed or feed.
c. Action that benefits another individual.
d. Honeybeee communication.
e. Affects neural circuitry in the hypothalamus.
f. Thinking behavior.
g. The study of animal behavior.
h. Defense of home range.

22.1   Approaches to the Study of Behavior (Page 440)

9. Two ways to explain behavior are _____ and _____ causation.

10. A behavior produced by forces of natural selection that makes the animal better adapted to its environment best fits
    a. proximate causation.
    b. ultimate causation.

22.2   Instinctive Behavioral Patterns (Page 441)

11. The male stickleback fish is very territorial. Bright red coloration during breeding season causes aggressive reaction behavior. The red color is best described as
    a. associative learning.
    b. sign stimulus.
    c. cognitive learning.
    d. none of the above.

22.3   Gene Effects on Behavior (Page 442)

12. The study of how genes determine behavior is called _____ .

22.4   How Animals Learn (Page 443)

True (T) or False (F) Question
If you believe the statement to be false, rewrite the statement as a true one.

13. Nonassociative learning requires an animal to form an association between stimulus and response.
    Answer: _____ Restatement: _____

    _____

14. Briefly define classical and operant conditioning.

22.5   Instinct and Learning Interact to Determine Behavior (Page 444)

15. Marler's work with white-crowned sparrows shows that
    a. the song template is learned only.
    b. the song template is genetically determined.
    c. birds have an instinctive program to learn the appropriate song.
    d. none of the above

22.6   Animal Cognition (Page 445)

True (T) or False (F) Question
If you believe the statement to be false, rewrite the statement as a true one.

16. Cognitive behavior can explain how the raven figures out how to obtain meat at the end of the string.
    Answer: _____ Restatement: _____

    _____

22.7   Behavioral Ecology (Page 446)

17. The study of how natural selection shapes behavior is called _____ .

22.8   The Cost-Benefit Analysis of Behavior (Pages 447)

18. Briefly discuss foraging and territorial behaviors in terms of maximum energy gains.

22.9   Migratory Behavior (Pages 448-449)

19. List several navigational aids animals employ in migration.

22.10   Reproductive Behaviors (Pages 450–451)

20. Polygyny refers to mating systems where
    a. a female mates with more than one male.
    b. a female mates with only one male.
    c. a male mates with more than one female.
    d. a male mates with only one female.
    e. none of the above

22.11 Communication Within Social Groups (Pages 452-453)

21. The "dance language controversy" arose from two very different areas of research regarding the waggle dance. Briefly describe the controversy.

22.12 Altruism and Group Living (Pages 454-455)

22. Selection that favors altruism directed toward relatives is called _____ _____.

22.13 Vertebrate Societies (Page 456)

23. Conflict in vertebrate societies centers on access to
    a. food.　　　　　b. mates.　　　　　c. work.　　　　　d. a and b are correct

22.14 Human Social Behavior (Pages 457-458)

24. Both _____ and _____ play key roles in human behavior.

Chapter Test

1. Which of the following terms does not belong when discussing migration?
    a. sun　　　　　b. stars　　　　　c. gravity　　　　　d. magnetic fields　　　　　e. cue recognition

2. Altruistic behavior is best represented by
    a. mate choice.
    b. complex courtship.
    c. optimal foraging.
    d. taking turns looking out for predators.
    e. all of the above

3. The waggle dance performed by honeybees communicates
    a. the direction of a food source.
    b. the distance of a food source.
    c. a and b are correct
    d. none of the above

4. The term most closely associated with ultimate causation is
    a. evolutionary forces.
    b. physiological changes.
    c. psychological pressures.
    d. climate fluctuation.
    e. none of the above

5. Egg retrieval behavior in geese is best explained by
    a. learned behavior.
    b. social behavior.
    c. innate behavior.
    d. operant conditioning.

6. Chemical signals used by many insects, fish, birds and mammals to alarm or attract other individuals are known as
    a. perfume.
    b. pheromones.
    c. foraging scents.
    d. none of the above

7. Shore crabs feed primarily on intermediate-size mussels rather than larger mussels that provide more energy but also require more energy to open. This behavior is best referred to as
    a. territorial.
    b. innate.
    c. optimal foraging.
    d. none of the above

Additional Study Help

Visit the ARIS (Assessment, Review, and Instruction System) site at aris.mhhe.com for quizzes, animations, and other study tools.

23 Planet Under Stress

Key Concepts Outline

Air and water pollution result mainly from increased industrialization, oil spills, the use of agricultural chemicals, and unwise disposal of wastes. (pages 462-465)
- Acid rain, precipitation that is polluted by sulfuric acid originating from factories, kills trees and lakes by lowering the pH level.
- Chlorofluorocarbons (CFCs) used in coolants, aerosols, and foaming agents are eating the earth's ozone, which exposes life on earth to harmful ultraviolet radiation.
- An increase in carbon dioxide in the atmosphere contributes to the greenhouse effect, which is responsible for increased global warming.

Habitat loss, overexploitation and introduced species are the main causes of the loss of biodiversity. (page 466)
Antipollution laws, pollution taxes, and economic evaluations are underway to reduce pollution. (pages 468-471)
- Nuclear power provides an alternative source of energy to the burning of coal and oil, although the challenges of operation must be overcome.
- The use of topsoil and groundwater, resources that took thousands of years to accumulate, must be decreased.
- With the destruction of tropical and temperate rain forests, organisms with potentially vital roles in the ecosystem are being lost.

The rapid growth of the human population is the underlying cause of many environmental concerns. (pages 472-475)
- Efforts are being made to slow the growth of the human population, but it's so large another 1-4 billion people could be added before stabilization occurs.

Environmental problems have been overcome through assessment, analysis, education, and commitment to making a difference. (pages 476-478)
- You can do your part by recycling, educating others about the environment, voting, and writing letters.

Key Terms Matching

1. _____ Biological magnification
2. _____ Acid rain
3. _____ Ozone hole
4. _____ CFCs
5. _____ Greenhouse effect
6. _____ Topsoil
7. _____ Groundwater
8. _____ Biodiversity
9. _____ Recycle
10. _____ Public education

a. Lost by repeated tilling.
b. Has led to an increase in lethal melanoma skin cancers.
c. Intensified by burning of fossil fuels.
d. This happens to toxins as they pass down the food chain.
e. To use over.
f. Richness of life.
g. A must if we are to learn how to save our resources.
h. Caused by sulfur in the upper atmosphere.
i. It is trapped in porous rocks.
j. Break down ozone.

23.1   Pollution (Page 462)

11. An interesting contradiction is that although the use of chlorinated hydrocarbons has been banned in the United States, they are
    a. still used in some medical applications.
    b. used when special permission can be obtained.
    c. still manufactured in the U.S. and exported to other countries.
    d. no longer a threat to the ecosystem.
    e. both a and d.

12. Biological magnification
    a. is the magnification of toxins in animal tissues as they pass through the food chain.
    b. is only a problem in invertebrate animals.
    c. causes producers in the food chain to die so there is a lack of food for animals higher in the food chain.
    d. is the result of inadequate phosphorus and carbon in the soil.
    e. causes thin shells in certain species of reptiles.

23.2   Acid Precipitation (Page 463)

True (T) or False (F) Questions
If you believe the statement to be false, rewrite the statement as a true one.

13. Acid rain is killing trees and changing the pH of many lakes so they no longer support life.
    Answer: _____ Restatement: _____
    _____

14. A problem with implementing solutions to control pollution is that they are expensive and no one is eager to foot the bill.
    Answer: _____ Restatement: _____
    _____

23.3   The Ozone Hole (Page 464)

15. A _____% drop in the atmospheric ozone content is estimated to lead to a _____% increase in the incidence of skin cancers.

23.4   The Greenhouse Effect (Page 465)

16. Increases in the amounts of greenhouse gases could increase global temperatures from:
    a. 1° to 4°.   b. 5° to 10°.   c. 5° to 12°.   d. 10° to 15°.   e. It will actually cause a reduction in temperature.

23.5   Loss of Biodiversity (Page 466)

17. Three factors that play a key role in many extinctions are _____, _____, and _____.

23.6   Reducing Pollution (Page 468)

True (T) or False (F) Questions
If you believe the statement to be false, rewrite the statement as a true one.

18. Laws and taxes designed to curb pollution will quickly correct the problem.
    Answer: _____ Restatement: _____
    _____

23.7   Finding Other Sources of Energy (Page 469)

19. Obstacles to the widespread use of nuclear power are _____, _____ and _____.
20. The building of nuclear power stations in the United States was dramatically slowed after _____.

23.8   Preserving Nonreplaceable Resources (Pages 470-471)

21. In the past _____ years about _____ of the world's rain forests have been burned to make pasture or have been cut for timber.

23.9   Curbing Population Growth (Pages 472-475)

22. The world population passed _____ people in 2004 and the annual increase is about _____ people.
    a. 6.5 million, 7.8 billion   d. 65 million, 7.8 million
    b. 6 billion, 9 million   e. 65 billion, 9 million
    c. 6.5 billion, 7.8 million

23.10   Preserving Endangered Species (Pages 476–478)

23. Discuss how each of the following help to preserve endangered species.
    a. Habitat restoration

    b. Captive propagation

    c. Conservation of ecosystems

23.11   Individuals Can Make the Difference (Page 479)

24. Discuss one thing you will implement to make a difference in your community.

Chapter Test

1. Which of the following can cause acid rain?
    a. soil erosion   d. clear-cutting forests
    b. industrial smokestacks   e. none of the above
    c. a decrease in the ozone layer

2. Widespread effects on the worldwide ecosystem are called
    a. disasters.   d. succession.
    b. global change.   e. both c and d.
    c. beneficial.

3. Why does it take so long for plastics to break down in nature?
    a. Ozone in the atmosphere slows down the process.
    b. Bacteria and fungi are unable to break down plastics.
    c. Plastics become stronger with age.
    d. Plastics can only be broken down by incineration.
    e. none of the above.

4. _____ released when coal is burned contributes to the production of acid rain.
    a. Nitrogen   b. Carbon   c. Sulfur   d. Phosphorus   e. Carbon dioxide

5. Lakes with a pH of _____ cannot support most life.
    a. 10   b. 8   c. 7   d. 5   e. none of the above

6. A serious drawback to the use of nuclear power is
    a. the disposal of radioactive materials.
    b. that it cannot provide very much energy.
    c. the shortage of radioactive materials.
    d. the difficulty in guarding nuclear power plants.
    e. both a and d.

7. To solve environmental problems we must
    a. assess the situation.
    b. educate the public.
    c. predict the consequences of environmental intervention.
    d. vote.
    e. all of the above.

Completion

8. An oil tanker called the _____ was involved in a huge oil spill in Alaska in 1989.
9. A class of chemicals called _____ are destroying the ozone layer.
10. The _____ effect prevents heat from radiating into space.
11. _____ helps prevent the waste of resources.
12. Plowing for the planting of crops has resulted in a loss of _____.
13. Water trapped beneath the soil in porous rock is called _____.
14. Humans first came to North America about _____ years ago.
15. An alarming trend in developing countries is the movement of humans to _____.

Additional Study Help

Visit the ARIS (Assessment, Review, and Instruction System) site at aris.mhhe.com for quizzes, animations, and other study tools.

24 The Animal Body and How It Moves

Key Concepts Outline

Four key transitions in animal body design created a great diversity of animal phyla. (page 484)
- These transitions include: radial to bilateral symmetry, no body cavity to body cavity, the development of segmented bodies, and protostome to deuterostome embryonic development.

Vertebrate bodies have different levels of organization. (pages 486-489)
- Tissues are groups of similar-type cells that perform a certain function.
- Organs are composed of different tissues that perform a certain function.
- Organ systems are groups of organs that work together to perform a specific function.

Adult tissues are grouped into four general classes: epithelial, connective, nervous and muscle tissue. (pages 490-497)
- The outermost of the principal tissues, epithelium, protects the tissues beneath it from dehydration.
- Connective tissue supports the body structurally, defends it with the immune system, and transfers materials via the blood.
- Bone is a dynamic connective tissue that undergoes remodeling in response to stress.
- Muscle tissue provides for movement and nerve tissue provides for regulation.

Animals' locomotion is accomplished through the force of muscles acting on the skeletal system of the animal. (pages 498-499)
- Many soft-bodied invertebrates are characterized by a hydraulic skeleton; an exoskeleton is characteristic of the arthropods.
- Echinoderms and vertebrates have endoskeletons to which muscles attach.

Muscle cells do the actual work of movement. (pages 500-503)
- Muscle cells contain large amounts of the protein filaments actin and myosin.
- Muscle cells contract when myosin "walks" along the actin filament, driven by the cleavage of ATP.

Key Terms Matching

1. _____ Tissues
2. _____ Organs
3. _____ Organ system
4. _____ Epithelium
5. _____ Connective tissue
6. _____ Fibroblasts
7. _____ Bone
8. _____ Adipose tissue
9. _____ Neurons
10. _____ Skeleton
11. _____ Haversian canal
12. _____ Tendons
13. _____ Myofibrils
14. _____ Actin filament
15. _____ Myosin filament
16. _____ Motor end plate
17. _____ Sarcoplasmic reticulum

a. Fat-accumulating cells.
b. This runs parallel to the length of the bone.
c. Types include loose, elastic and dense.
d. One end is a long rod, the other has a double-headed globular region.
e. Several tissues grouped together.
f. Made of myofilaments.
g. Where a neuron almost touches a muscle cell.
h. A group of the same type of cells.
i. One end is anchored at the Z line.
j. Where muscle fibers store Ca^{++}.
k. A group of organs working together.
l. Glial cells supply them with nutrients.
m. One type lines the lungs and major cavities of the body.
n. Support and protection of organs.
o. Used to attach muscles to bones.
p. This includes blood and fat cells.
q. Collagen fibers coated with a calcium phosphate salt.

24.1   Innovations in Body Design (Pages 484-485)

18. A profound difference between the sponges and the other animals is that
    a. sponges completely lack tissues.
    b. sponges have radial symmetry while other animals have bilateral symmetry.
    c. cephalization is seen in sponges but not in other animals.
    d. organ systems in sponges work differently than in other animals.
    e. both b and d

19. _____ exhibit radial symmetry.
    a. Copepods   b. Cnidarians   c. Ctenophorans   d. Planarians   e. Both b and c

20. _____ Acoelomate   a. Mollusks and arthropods
21. _____ Pseudocoelomate   b. Flatworms
22. _____ Coelomate   c. Nematodes

23. Segmentation may represent an adaptation for _____.
24. Segmentation was a powerful means of _____ the animal body.
25. Deuterostome embryos undergo _____ cleavage and the blastopore becomes the _____.
26. Protostome embryos undergo _____ cleavage and the blastopore becomes the _____.

24.2   Organization of the Vertebrate Body (Pages 486-489)

27. Which of the following systems coordinates and integrates the body's activities?
    a. skeletal   b. circulatory   c. endocrine   d. nervous

28. Which of the following systems removes metabolic wastes from the bloodstream?
    a. integumentary   b. urinary   c. digestive   d. respiratory   e. lymphatic/immune

24.3   Epithelium Is Protective Tissue (Pages 490-491)

29. _____ Simple squamous   a. Lines the stomach.
30. _____ Cuboidal   b. Secretes mucus, many cilia.
31. _____ Simple columnar   c. Lines the mouth.
32. _____ Stratified squamous   d. Lines the lungs.
33. _____ Stratified columnar   e. Covers ovaries.

24.4   Connective Tissue Supports the Body (Pages 492-494)

34. _____ Fibroblasts   a. Loose, dense or elastic. Ligaments and tendons.
35. _____ Macrophages   b. Oxygen transport.
36. _____ Chondrocytes   c. Shock absorption.
37. _____ Erythrocytes   d. Attack invading microbes.

24.5   Muscle Tissue Lets the Body Move (Pages 495-496)

38. The three types of muscle cells are _____, _____ and _____.
39. In cardiac muscle, microfilaments are bunched into fibers called _____.

24.6   Nerve Tissue Conducts Signals Rapidly (Page 497)

40. Label the parts of the neuron (a – d).

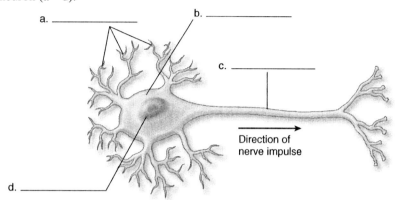

a. _____   b. _____

c. _____

Direction of
nerve impulse

d. _____

24.7   Types of Skeletons (Pages 498-499)

41. The human skeleton is made of _____ bones. There are _____ bones in the axial skeleton and _____ in the appendicular skeleton.

42. _____ Endoskeleton   a. Fluid-filled cavity.
43. _____ Exoskeleton   b. Chitenous covering.
44. _____ Hydraulic skeleton   c. Muscles attach to bones.

24.8   Muscles and How They Work (Pages 500-503)

45. Two primary proteins involved in muscle contraction are _____ and _____.

46. When Ca^{++} concentration is low, the muscle is
    a. relaxed.   b. contracted.

47. Ca^{++} is stored in
    a. the mitochondria.
    b. the nucleus.
    c. the sarcoplasmic reticulum.
    d. none of the above

Chapter Test

1. Epithelial tissue arises from embryonic
    a. ectoderm.   b. mesoderm.   c. endoderm.   d. neural tube cells.   e. fluid.

2. Epithelial tissue functions to
    a. secrete a variety of materials.   d. all of the above.
    b. protect other tissues.   e. both a and b.
    c. provide sensory surfaces.

3. Hormones are secreted by what type of tissue?
    a. connective   b. nervous   c. muscle   d. epithelial   e. skeletal

4. Macrophages originate from which embryonic tissue?
   a. ectoderm   b. mesoderm   c. endoderm   d. neural tube   e. none of the above

5. What type of connective tissue is made of collagen fibers that are coated with calcium phosphate salt?
   a. adipose   b. fibroblasts   c. bone   d. cartilage   e. lymphocytes

6. _____ function(s) to attack cells that are infected with viruses.
   a. Adipose   b. Fibroblasts   ·   c. Bone   d. Cartilage   e. Lymphocytes

7. Myofibrils are
   a. the basic unit of the muscle cell.   d. only found in cardiac muscle.
   b. microfilaments bunched into fibers.   e. necessary for muscle contraction.
   c. only found in skeletal muscle.

8. Glial cells
   a. transmit nerve impulses.   d. supply neurons with nutrients.
   b. span gaps between neurons.   e. act as antennae for neurons.
   c. secrete neurotransmitters.

9. The human body has a spine that is made up of _____ vertebrae. There are _____ pairs of ribs that curve forward from the vertebrae.
   a. 25, 10   b. 33, 12   c. 41, 10   d. 50, 12   e. 52, 10

10. Osteoblasts
    a. make up Haversian canals in bones.   d. lose their nucleus upon reaching maturity.
    b. produce the bone marrow.   e. are found in reptiles but not in mammals.
    c. secrete collagen.

Matching

11. _____ Erythrocytes   a. Cells that engulf and digest invading microorganisms.
12. _____ Lymphocytes   b. Cells that transport oxygen and carbon dioxide in the body.
13. _____ Plasma   c. Antibodies are produced by these cells.
14. _____ Fibroblasts   d. This is the fluid in which red blood cells are suspended.
15. _____ Macrophages   e. These cells secrete proteins into the spaces between cells.

Additional Study Help

Visit the ARIS (Assessment, Review, and Instruction System) site at aris.mhhe.com for quizzes, animations, and other study tools.

25 Circulation

Key Concepts Outline

Vertebrate circulatory systems perform three general functions: transport, regulation and protection. (pages 508-509)
- The two main kinds of circulatory systems are open or closed; vertebrates have a closed circulatory system, in which the blood stays within the vessels.

A variety of types of vessels carry blood through the body. (pages 510-512)
- Arteries and arterioles carry blood away from the heart; the exchange of gases, food and waste occurs from capillaries; venules and veins return blood to the heart.

The lymphatic system collects interstitial fluid and returns it to the circulatory system. (page 513)
- Three other functions of the lymphatic system are: return proteins to circulation, transport fats absorbed after digestion, and assist with the body's defence.

Blood is composed of fluid called plasma and different kinds of cells. (page 514-515)
- Erythrocytes (red blood cells) carry oxygen; leukocytes (white blood cells) defend the body; platelets assist in blood clotting.

In fish, blood from the heart goes to the gills and then to the rest of the body before returning to the heart. (page 516)
- A fish heart is basically a tube with four chambers, two for collection and two for pumping.

Amphibians and reptiles have three-chambered hearts with two atria and one ventricle. (page 517)
- Pulmonary circulation transports blood to and from the lungs; systemic circulation transports blood to and from the rest of the body.
- Reptiles have a septum that partially divides the ventricle.

Mammals, birds and crocodiles have four-chambered hearts with two atria and two ventricles that keep deoxygenated and oxygenated blood completely separated, which supports the high metabolism of these animals. (pages 518-520)
- The contraction of the heart starts in the right atrium and spreads as a wave across the heart.
- Heart performance can be measured by a variety of methods including EKG and blood pressure.

Humans may suffer from a variety of heart diseases that are often associated with fatty materials accumulating in the arteries. (page 521)

Key Terms Matching

1. _____ Gastrovascular cavity
2. _____ Open circulatory system
3. _____ Arteries
4. _____ Capillaries
5. _____ Veins
6. _____ Lymphatic system
7. _____ Plasma
8. _____ Erythrocytes
9. _____ Platelets
10. _____ Atrium
11. _____ Ventricle
12. _____ Pulmonary circulation
13. _____ Pulmonary arteries
14. _____ Tricuspid valve
15. _____ Semilunar valve
16. _____ Sinoatrial (SA) node
17. _____ Systolic pressure
18. _____ Atherosclerosis

a. A second circulatory system.
b. Blood flow to and from the lungs.
c. High pressure associated with ventricular contraction.
d. Fluid component of blood.
e. Carry deoxygenated blood to the lungs.
f. Performs digestion and transport for some invertebrates.
g. Blood leaves the heart through these.
h. Collection chamber.
i. Located between the left ventricle and the aorta.
j. Collect blood and carry it back to the heart.
k. Muscular chamber that pumps blood.
l. Smallest diameter of all blood vessels.
m. Cell fragments with no nuclei that aid in clotting.
n. Arterial passageways are narrowed.
o. Flattened disks with central depressions on each side.
p. The site where heartbeats originate.
q. Prevents backflow of blood from the right ventricle.
r. Characteristic of mollusks and arthropods.

25.1   Open and Closed Circulatory Systems (Pages 508-509)

19. Hemolymph is
    a. a fluid containing red blood cells that is characteristic of amphibians.
    b. fluid found in insects consisting of blood and interstitial fluid.
    c. the interstitial fluid of mammals.
    d. a lighter-weight fluid characteristic of many bird groups.
    e. none of the above.

20. The circulatory system and endocrine system are interrelated in which of the following ways?
    a. The glands of the endocrine system produce and secrete hormones that are delivered to other body sites via the circulatory system.
    b. The temperature of blood regulates the rate at which many endocrine glands produce hormones.
    c. Hormones produced by the endocrine system are absorbed by lymphocytes and delivered to other body sites.
    d. Hormones produced by the endocrine system control the activities of blood cells.
    e. The circulatory system is a part of the endocrine system.

25.2   Architecture of the Vertebrate Circulatory System (Pages 510-512)

21. Arrange the following in order from largest to smallest:
    a. Capillaries, arteries, arterioles.
    b. Capillaries, arterioles, arteries.
    c. Arterioles, arteries, capillaries.
    d. Arteries, arterioles, capillaries.
    e. Arteries, capillaries, arterioles.

22. The rate of blood flow in humans is about
    a. 5 pints per minute.
    b. 5 liters per minute.
    c. 5 gallons per minute.
    d. 10 liters per minute.
    e. 10 gallons per minute.

25.3   The Lymphatic System: Recovering Lost Fluid (Page 513)

23. Inside each lymph node are _____ that function to _____.
24. Lymph is moved through the lymph system by _____.

25.4   Blood (Pages 514-516)

25. _____ B-lymphocyte   a. Inflammation.
26. _____ Monocyte   b. Transport carbon dioxide.
27. _____ Basophil   c. Produce antibody molecules.
28. _____ Erythrocyte   d. Large nucleus, functions in immunity.

25.5   Fish Circulation (Page 516)

29. In fish, blood flows in which direction?
    a. conus arteriosus, sinus venosus, atrium, ventricle
    b. ventricle, atrium, conus arteriosus, sinus venosus
    c. sinus venosus, atrium, ventricle, conus arteriosus
    d. ventricle, conus arteriosus, atrium, sinus venosus
    e. none of the above

25.6   Amphibian and Reptile Circulation (Page 517)

30. Amphibians supplement oxygenation of their blood through a process known as _____.
31. A circulatory system in which blood moves from the heart to the rest of the body is known as _____.

25.7   Mammalian and Bird Circulation (Pages 518-520)

32. _____ Bicuspid valve   a. Carries high-pressure blood to the body.
33. _____ Pulmonary veins   b. Transports blood back to the right atrium.
34. _____ Aorta   c. Located between the left atrium and ventricle.
35. _____ Vena cava   d. Transport oxygenated blood to the left atrium.

36. In normal blood pressures the _____ rate averages 110 – 130 while the _____ rate is about 70 – 90.
37. Heart performance can be monitored by _____, _____, and _____.

25.8   Cardiovascular Diseases (Page 521)

38. _____ Arteriosclerosis   a. Hardening of the arteries.
39. _____ Angina pectoris   b. Interference of blood supply to the brain.
40. _____ Stroke   c. "Chest pain."

Chapter Test

1. Blood leaves the heart through
   a. veins.   b. capillaries   c. arteries.   d. venules.   e. arterioles.

2. Blood moves from capillaries into
   a. veins.   b. subcapillaries.   c. arteries.   d. venules.   e. arterioles.

3. _____ carry blood back to the heart.
   a. Veins   b. Capillaries   c. Arteries   d. Venules   e. Arterioles

4. An important characteristic of an artery is
   a. flexibility.   d. its affinity for carbon monoxide.
   b. its radius of about 8 micrometers.   e. both a and b.
   c. its ability to return blood to the heart.

5. Arterioles differ from arteries in that
   a. they are smaller in diameter.
   b. they are larger in diameter.
   c. hormones can cause the muscle surrounding them to relax and they can enlarge in diameter.
   d. both a and c.
   e. both b and c.

6. The thin walls of capillaries are important because they
   a. allow capillaries to easily change diameter.
   b. allow for the transport of gases and metabolites.
   c. must be flexible to accommodate surges of blood.
   d. must withstand high pressure.
   e. none of the above.

7. The lymphatic system functions to
    a. filter bacteria and debris from lymph fluid.
    b. remove carbon dioxide from the blood.
    c. add oxygen to the blood.
    d. adjust the water content of plasma.
    e. add bicarbonate to the blood.

8. The lymphatic system
    a. transports fats absorbed from the intestine.
    b. returns proteins to circulation.
    c. filters debris from lymph fluid.
    d. destroys debris removed from lymph fluid.
    e. all of the above.

9. A function of proteins dissolved in blood plasma is to
    a. provide energy to blood cells.
    b. prevent bacteria from surviving in the bloodstream.
    c. prevent the osmotic loss of water from the plasma.
    d. prevent salts from building up in the blood.
    e. maintain a constant pH in the blood.

10. Heart contraction is initiated by
    a. the superior vena cava.
    b. Purkinje fibers.
    c. an electrical impulse of the appropriate intensity.
    d. the sinoatrial node.
    e. the bundle of His.

11. During a routine examination of your elderly dog, the veterinarian tells you that she has a slight heart murmur. A likely cause of the murmur is
    a. the heart valves encountering fat accumulation when they close.
    b. turbulence in the heart caused by valves opening or closing incompletely.
    c. failure of the heart to completely empty.
    d. weakening of the cardiac muscle.
    e. both a and d.

Matching

12. _____ Megakaryocytes   a. Bits of cytoplasm pinch off of these cells.
13. _____ Hematocrit   b. They are filled with hemoglobin.
14. _____ Erythrocytes   c. Fibrin fibers cause these cells to stick together and assist in clotting.
15. _____ Leukocytes   d. These cells can migrate out of the bloodstream.
16. _____ Platelets   e. This refers to the volume of blood that is occupied by cells.

Additional Study Help

Visit the ARIS (Assessment, Review, and Instruction System) site at aris.mhhe.com for quizzes, animations, and other study tools.

26   Respiration

Key Concepts Outline

Respiration is the uptake of oxygen gas from air and the simultaneous release of carbon dioxide. (page 526)
- Marine and aquatic vertebrates extract oxygen from water with gills.
- Terrestrial vertebrates use lungs to exchange gases with the blood.

Countercurrent flow enables fish gills to efficiently extract oxygen from water. (page 527)
- Blood always encounters water with a higher oxygen concentration during countercurrent flow.

Lungs enable terrestrial animals to extract oxygen from the air. (pages 528-529)
- Mammals' lungs contain alveoli that create a large surface area for diffusion.
- Birds also have air sacs that allow them to perform a crosscurrent flow of gases, making their lung the most efficient of all.

In mammals, lungs are located in the thoracic cavity, the volume of which increases or decreases when the diaphragm contracts or relaxes. (pages 530-531)
- During inhalation air is drawn into the lungs when the diaphragm contracts and the rib cage moves out to expand the chest cavity; during exhalation air is pushed out when the ribs and diaphragm return to their resting position.
- Vital capacity is the extra amount of air that can be forced into and out of the lungs.

Oxygen is transported by the protein hemoglobin in red blood cells; most carbon dioxide is transported dissolved in the cytoplasm of red blood cells. (pages 532-533)
- The formation of bicarbonate ions helps buffer the pH of the blood plasma.
- Nitric oxide is also transported by hemoglobin.

Cancer is caused by damage to growth-regulating genes. (pages 534-535)
- Cigarette smoking is the principal cause of lung cancer.
- Smoking produces lung cancer by introducing carcinogens into the lungs.

Key Terms Matching

1. _____ Gill
2. _____ Spiracles
3. _____ Lung
4. _____ Countercurrent flow
5. _____ Alveoli
6. _____ Bronchiole
7. _____ Thoracic cavity
8. _____ Trachea
9. _____ Diaphragm
10. _____ Pleural membrane
11. _____ Vital capacity
12. _____ Hemoglobin
13. _____ Bicarbonate
14. _____ Lung cancer
15. _____ p53 protein

a. Ion that helps buffer blood plasma.
b. Short passageway connecting alveoli and a main air sac.
c. Occurs in animals with gills.
d. Area in the chest.
e. Openings of tracheae to the environment.
f. Respiratory organ of terrestrial vertebrates.
g. Protein that transports oxygen.
h. Small chambers like "clusters of grapes."
i. A leading cause of adults' deaths.
j. Muscle separating the thoracic and abdominal cavities.
k. Respiratory organ found in marine invertebrates and fish.
l. Lining covering the lungs.
m. A tumor suppressor.
n. Extra amount of air that can be forced into and out of the lungs.
o. Passageway for air in mammals.

26.1   Types of Respiratory Systems (Page 526)

16. Insects obtain oxygen through
    a. tracheae.   b. gills.   c. water.   d. lungs.   e. book lungs.

26.2   Respiration in Aquatic Vertebrates (Page 527)

17. The gills of the bony fishes are considered the most efficient respiratory machines ever evolved because they
    a. require so little oxygen.
    b. use reverse osmosis to extract oxygen from water.
    c. concentrate oxygen so efficiently that the blood in gills may have twice the oxygen as the surrounding water.
    d. are able to extract up to 85% of the available oxygen from water.
    e. both a and b are true.

26.3   Respiration in Terrestrial Vertebrates (Pages 528-529)

True (T) or False (F) Questions
If you believe the statement to be false, rewrite the statement as a true one.

18. Bird lungs, like the gills of fish, use a countercurrent flow to oxygenate blood.
    Answer: _____ Restatement: _____
    _____

19. Air flows through the lungs of birds in one direction, from front to back.
    Answer: _____ Restatement: _____
    _____

26.4   The Mammalian Respiratory System (Pages 530-531)

20. The pleural membrane is
    a. used by fishes to achieve countercurrent flow and oxygenate their blood.
    b. a smooth membrane covering each lung.
    c. what allows birds to achieve such high oxygen concentrations in their blood.
    d. divided into many tiny cavities called bronchi.
    e. only seen in aquatic mammals.

21. The thoracic cavity
    a. functions to remove carbon dioxide from the blood in mammals.
    b. is divided into many tiny cavities called bronchi.
    c. is filled with interpleural fluid.
    d. contains the lungs.
    e. is a characteristic of the amphibians.

26.5   How Respiration Works: Gas Exchange (Pages 532-533)

22. Hemoglobin molecules are able to carry
    a. oxygen.
    b. carbon dioxide.
    c. nitric oxide.
    d. all of the above.
    e. a and c only.

23. The majority of carbon dioxide is removed from the body by
    a. binding to hemoglobin in red blood cells.
    b. binding to carrier proteins in lymphocytes.
    c. the alveoli.
    d. diffusing from the skin.
    e. dissolving in the cytoplasm of red blood cells.

26.6   The Nature of Lung Cancer (Pages 534-535)

24. When the Rb protein in cells is disabled the result is _____.
25. The function of p53 is to inspect the cell for _____.
26. The best way to avoid lung cancer is by _____.

Chapter Test

1. Extending from spiracles are networks of air ducts called _____ that deliver air to all body parts of terrestrial arthropods.
    a. alveoli   b. tracheae   c. arterioles   d. capillaries   e. bronchioles

2. Countercurrent flow enables _____ to extract oxygen from water very efficiently.
    a. alveoli   b. tracheae   c. capillaries   d. fish gills   e. bird air sacs

3. The surface area available for diffusion in lungs is increased by the presence of
    a. alveoli.   b. cilia.   c. air sacs.   d. villi.   e. papillae.

4. The _____ of birds enable them to get more oxygen from air than mammals, which supports their ability to fly.
    a. spiracles   b. gills   c. air sacs   d. lungs   e. diaphragm

5. The lungs are covered by a thin, smooth membrane called the
    a. pericardium.   b. mesentery.   c. renal capsule.   d. pleura.   e. cortex.

6. The volume of air that remains in the lungs after a maximum expiration is called the
    a. residual volume.   b. vital capacity.   c. tidal volume.   d. expiratory reserve.   e. inspiratory reserve.

7. During inhalation the
    a. lungs expand in size.
    b. diaphragm contracts.
    c. volume of the thoracic cavity increases.
    d. rib cage moves outwards and upwards.
    e. all of these.

8. Alveoli are connected to the bronchi by
    a. blood vessels.   b. connective tissue.   c. muscle fibers.   d. the pleural membrane.   e. bronchioles.

9. The effect of carbon dioxide on hemoglobin is that
    a. it causes the hemoglobin to change shape and unload the oxygen it is carrying.
    b. it changes its shape so that hemoglobin can no longer bind to oxygen.
    c. it increases hemoglobin's affinity for oxygen.
    d. hemoglobin is denatured.
    e. both a and d.

10. Most carbon dioxide is transported
    a. by hemoglobin.   b. dissolved in plasma.   c. by platelets.
    d. dissolved in red blood cells' cytoplasm.   e. by white blood cells

11. Oxygen binds to the _____ part of hemoglobin.
    a. copper   b. iron   c. sodium   d. amino acid   e. iodine

12. In order for lung cancer to be initiated,
    a. several thousand genes must be mutated.
    b. only a few critical genes need to be mutated.
    c. lung cells must be exposed to some kind of radiation.
    d. the patient must be in poor health.
    e. none of the above.

Additional Study Help

Visit the ARIS (Assessment, Review, and Instruction System) site at aris.mhhe.com for quizzes, animations, and other study tools.

27 The Path of Food Through the Animal Body

Key Concepts Outline

Ingested calories are either metabolized by the body or stored as fat. (pages 540-542)
- Healthy diets contain those necessary substances that animals cannot manufacture, such as essential amino acids, vitamins, and trace elements.

In most animals the digestion of food is done by a gastrovascular cavity or tubular digestive tract. (page 544)
- A digestive tract with a separate mouth and anus allows for specialization of regions for different functions.

Various modifications of a tubular gastrointestinal tract composed of different layers of tissues exist in vertebrates. (page 545)
- Smooth muscle of the GI tract pushes food through the tract.

Food is broken down into smaller pieces in the mouth, and the addition of saliva begins chemical digestion. (pages 546-547)
- Salivary amylase begins the digestion of starch.

Food is propelled to the stomach by the esophagus. (pages 548-549)
- The chemical digestion of proteins by pepsin and HCl occurs in the stomach.

Most chemical digestion and the absorption of the products of digestion occur in the small intestine; the large intestine concentrates undigested material. (pages 550-551)
- Many enzymes that perform chemical digestion in the small intestine are made by the pancreas.
- Villi create a lot of surface area for the absorption of the products of chemical digestion.

The digestive systems of different mammals are reflective of the animals' diets. (pages 552-553)
- Cellulose is digested by microorganisms in the rumen or cecum of many animals and vitamin K is synthesized by intestinal bacteria.

Chemical digestion depends on secretions from the pancreas and liver. (page 554)
- Enzymes and bicarbonate are made by the pancreas; bile is made by the liver.

Key Terms Matching

1. _____ Calories
2. _____ Essential amino acids
3. _____ Trace elements
4. _____ Gastrovascular cavity
5. _____ Chemical digestion
6. _____ Ruminant
7. _____ Cecum
8. _____ Mastication
9. _____ Peristalsis
10. _____ Sphincter
11. _____ Chyme
12. _____ Duodenum
13. _____ Villi
14. _____ Pancreas
15. _____ Liver
16. _____ Bile

a. Tiny, fingerlike projections lining intestine.
b. Carbohydrates have 4.1 of these per gram on average.
c. Muscular contractions that allow us to swallow when upside down.
d. Accessory organ that produces enzymes and bicarbonate.
e. Chewing.
f. Performs digestion and distribution of nutrients.
g. Ring of smooth muscle that controls food movement.
h. Obtained from proteins we eat.
i. Partially digested food and gastric juice mixture.
j. Creates lots of surface area for lipase activity.
k. First portion of the small intestine.
l. Essential minerals.
m. The largest internal organ of the body.
n. Pouch located at the start of the large intestine.
o. Breaks large food molecules into smaller subunits.
p. Herbivore with a four-chambered stomach.

27.1   Food for Energy and Growth (Pages 540–542)

17. According to the USDA a healthy diet should include 2 – 3 servings a day of/from the
    a. fats and sweets.   b. milk group.   c. fruit group.   d. vegetable group.   e. bread and cereals.

18. According to the USDA a healthy diet should include sparing use of/from the
    a. fats and sweets.   b. milk group.   c. fruit group.   d. vegetable group.   e. bread and cereals.

19. Individuals with a Body Mass Index value of _____ or over are considered overweight.
    a. 15   b. 20   c. 25   d. 30   e. 50

Match the vitamin with the correct deficiency syndrome.
20. _____ Vitamin B_{12} a. Scurvy.
21. _____ Folic acid   b. Severe bleeding.
22. _____ Vitamin A   c. Night blindness.
23. _____ Vitamin C   d. Anemia, diarrhea.
24. _____ Vitamin K   e. Pernicious anemia.

27.2   Types of Digestive Systems (Page 544)

25. Unicellular organisms digest their food _____ while multicellular animals digest their food
    _____ in a(n) _____.

27.3   Vertebrate Digestive Systems (Page 545)

26. Rabbits, horses and other herbivores use a bacteria-filled pouch called the _____ to digest _____.

27.4   The Mouth and Teeth (Pages 546-547)

27. Label the vertebrate dentition (a – d).

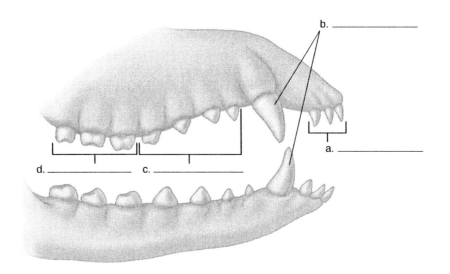

27.5   The Esophagus and Stomach (Pages 548-549)

True (T) or False (F) Questions
If you believe the statement to be false, rewrite the statement as a true one.

28. Carbohydrates and fats are not significantly digested in the stomach.
    Answer: _____ Restatement: _____

    _____

29. The entrance of food into the duodenum is controlled by the pyloric sphincter.
    Answer: _____ Restatement: _____

    _____

27.6   The Small and Large Intestines (Pages 550-551)

30. The majority of the small intestine consists of the
    a. duodenum.   b. ileum.   c. cecum.   d. colon.   e. villus.

31. The main function of the large intestine is
    a. the breakdown and absorption of fats.
    b. the breakdown and absorption of proteins.
    c. the breakdown of plant fiber and cellulose.
    d. concentration of solid wastes.
    e. emulsification of vitamins.

27.7   Variations in Vertebrate Digestive Systems (Pages 552-553)

32. The omasum and abomasum are a characteristic of
    a. all vertebrates.   b. amphibians.   c. ruminants.   d. birds.   e. both b and d

27.8   Accessory Digestive Organs (Pages 554-555)

True (T) or False (F) Questions
If you believe the statement to be false, rewrite the statement as a true one.

33. Edema may indicate a drop in plasma proteins due to malfunctioning of the kidneys.
    Answer: _____ Restatement: _____

    _____

34. The arrival of fatty foods in the duodenum triggers the release of bile from the gallbladder.
    Answer: _____ Restatement: _____

    _____

Chapter Test

1. Fats have a higher energy content per gram than do carbohydrates because they
   a. are more easily converted into ATP.
   b. are more easily metabolized.
   c. have more energy-rich bonds.
   d. release their energy at a lower temperature.
   e. none of the above.

2. Fats are a necessary part of the diet to
   a. provide energy.
   b. manufacture cell membranes.
   c. insulate nervous tissues.
   d. build a variety of structures in the cell.
   e. all of the above.

3. A diet that is high in fiber is important because
   a. the bulk helps to move food through the colon at a faster rate.
   b. fiber is a source of essential amino acids.
   c. fiber is an important source of nutrients.
   d. fiber is an important source of vitamins.
   e. the majority of protein in foods is in the bulk.

4. A consequence of a diet lacking vitamin C is
   a. the inability to synthesize hemoglobin.
   b. the development of diabetes.
   c. the development of scurvy.
   d. memory loss.
   e. the inability to break down carbohydrates.

5. Trace elements are
   a. only necessary for plant growth.
   b. obtained from the plants that we consume in our diet.
   c. obtained from the animal products in our diet.
   d. both a and b.
   e. both b and c.

6. _____ is(are) necessary to break up proteins, lipids and other nutrients to release energy.
   a. High temperatures   b. Acids   c. Enzymes   d. Trace elements   e. Oxygen

7. Chyme is
   a. the enzyme and acid mixture that breaks down food in the stomach.
   b. the name used for the infoldings of epithelial tissue inside of the stomach.
   c. the name of the sphincter located at the end of the esophagus.
   d. a mixture of digestive juices and food.
   e. the mucosa that lines the inside of the stomach and protects it from the acids located there.

8. Gastrin functions to
   a. inhibit the production of HCl in the stomach.
   b. regulate the synthesis of HCl in the stomach.
   c. absorb fats in the small intestine.
   d. regulate water loss from the large intestine.
   e. kill any bacteria that enter the stomach.

9. Aspirin can cause an upset stomach because it
   a. increases acid production.
   b. decreases acid production and slows digestion.
   c. is absorbed through the lining of the stomach.
   d. slows the breakdown of fats.
   e. increases bacterial multiplication in the stomach.

10. Gastric pits
    a. are lesions in the mucosa of the stomach.
    b. are invaginations in the stomach epithelium.
    c. are the site of actual digestion in the stomach.
    d. in the small intestine produce digestive enzymes.
    e. are not acquired until middle age and are due to a diet low in fiber.

Matching

11. _____ Small intestine
12. _____ Stomach
13. _____ Duodenum
14. _____ Large intestine
15. _____ Liver

a. Sodium and vitamin K are absorbed through this structure, and it acts as a refuse dump for undigested food.
b. Bicarbonate and digestive enzymes are delivered to this site.
c. Bile salts are produced by this structure.
d. Villi and microvilli increase the absorptive surface of this site.
e. Gastrin is produced here.

Additional Study Help

Visit the ARIS (Assessment, Review, and Instruction System) site at aris.mhhe.com for quizzes, animations, and other study tools.

28 Maintaining the Internal Environment

Key Concepts Outline

Negative feedback loops inhibit reactions once a desired product has been made, and they have an important role in maintaining homeostasis. (pages 560-561)
- Body temperature and blood glucose are kept within an acceptable range with negative feedback.

Among the most important functions of osmoregulatory organs are to eliminate the nitrogen by-products of protein metabolism and maintain water balance. (pages 562-563)
- Organisms have a variety of structures that filter fluids and waste products and reabsorb water.

There are millions of disposal units called nephrons in the kidney of a freshwater fish; the basic design and function of this nephron is observed in terrestrial vertebrates. (pages 564-567)
- Birds and mammals are able to reabsorb water and produce a hypertonic urine with some modifications to the loop of Henle.

Mammalian kidneys filter waste from the blood and reabsorb water, nutrients and ions before excreting urine. (pages 568-569)
- Other structures of the mammalian urinary system are ureters, urinary bladder, and urethra.
- The five steps to forming urine are: pressure filtration, water reabsorption, selective reabsorption of nutrients and ions, tubular secretion and additional water reabsorption.

The metabolism of amino acids and nucleic acids produces ammonia. (page 571)
- Bony fish excrete ammonia, but other vertebrates produce urea or uric acid that are less toxic and able to be excreted with less water.

Key Terms Matching

1. _____ Negative feedback loop		a. Excretory structures of earthworms.
2. _____ Osmoregulation		b. Process that adds substances to the urine in the ascending loop.
3. _____ Nephridia		c. Transports urine from a kidney to the bladder.
4. _____ Malpighian tubules		d. Nitrogenous wastes are converted to this by mammals.
5. _____ Kidney		e. Stimulates the production of concentrated urine.
6. _____ Nephron		f. Excretory organs found in insects.
7. _____ Glomerulus		g. Location of water reabsorption.
8. _____ Loop of Henle		h. Contains water, urea, glucose, amino acids and various ions.
9. _____ Ureter		i. Excreted by bony fish.
10. _____ Urethra		j. There are thousands of these disposal units.
11. _____ Glomerular filtrate		k. Control system that stops a reaction from occurring.
12. _____ Tubular secretion		l. Excretory organ of vertebrates.
13. _____ Antidiuretic hormone		m. Transports urine out of the body.
14. _____ Ammonia		n. Network of capillaries that acts as a filtration device.
15. _____ Urea		o. Nitrogenous waste made by birds and reptiles.
16. _____ Uric acid		p. Regulation of salt and water in an organism.

28.1 How the Animal Body Maintains Homeostasis (Pages 560–561)

17. Body temperature changes are detected by the _____.
18. _____ are ectothermic and attempt to maintain a constant body temperature by basking in the sun.
19. Insulin _____ blood glucose.

28.2   Regulating the Body's Water Content (Pages 562–563)

20. Flatworms release fluids and wastes from their bodies through tubules called _____ which branch into

    _____.

21. _____ are extensions of the digestive tract of insects that collect water and wastes.

28.3   Evolution of the Vertebrate Kidney (Pages 564–567)

True (T) or False (F) Questions
If you believe the statement to be false, rewrite the statement as a true one.

22. Organisms that depend on water from food and metabolic processes produce strongly hypertonic urine.
    Answer: _____   Restatement: _____

    _____

23. Amphibians absorb sodium across their skin to compensate for its loss in their urine.
    Answer: _____   Restatement: _____

    _____

28.4   The Mammalian Kidney (Pages 568–570)

24. Label the parts of the kidney (a – e).

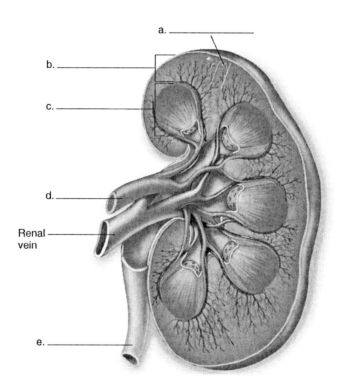

a. _____

b. _____

c. _____

d. _____

Renal vein

e. _____

28.5   Eliminating Nitrogenous Wastes (Page 571)

25. When _____ and _____ are metabolized, nitrogenous waste is produced and must be eliminated
    from the body.
26. Urea is produced by the _____.
27. Bony fish excrete _____ by diffusion through the gills or in a very dilute urine.

Chapter Test

1. Which of these is NOT correctly matched with its function?
   a. ureters – transport urine from the kidney to the bladder
   b. renal pelvis – collects urine from renal tissue
   c. kidneys – produce urea
   d. urethra – transports urine out of the body
   e. urinary bladder – stores urine

2. Which of these is NOT found in the glomerular filtrate of a healthy individual?
   a. water
   b. glucose
   c. H^+, Na^+, Ca^{++}
   d. white blood cells
   e. urea

3. Antidiuretic hormone
   a. is made by the hypothalamus.
   b. causes the distal tubules and collecting ducts to absorb more water.
   c. is produced when you become dehydrated.
   d. causes the production of concentrated urine.
   e. all of the above.

4. The production of _____ requires considerable energy, which is energy well spent since embryos developing in shelled eggs must live with this nitrogenous waste in the egg.
   a. ammonia
   b. urine
   c. uric acid
   d. urea
   e. aldosterone

5. Which of these is an example of negative feedback?
   a. heart rate and blood pressure increase during the sympathetic nervous system response
   b. hypothalamic stimulation of shivering stops once body temperature increases
   c. more water is reabsorbed when an individual is dehydrated
   d. oxytocin stimulates milk releases and its production continues while an infant is being breast fed
   e. pepsin digests proteins in the stomach when the pH is acidic

6. The body fluids of a marine bony fish are _____ to the surrounding seawater, which causes water to leave their bodies by diffusion across their gills.
   a. isotonic   b. hypotonic   c. hypertonic

7. Contractile vacuoles are used by _____ to expel metabolic wastes.
   a. earthworms
   b. amphibians
   c. insects
   d. *Paramecium*
   e. flatworms

8. Kidneys of _____ are the most efficient at conserving water.
   a. humans
   b. desert mammals
   c. marine mammals
   d. terrestrial reptiles
   e. freshwater fish

9. Blood to be filtered arrives at the kidneys via the
   a. renal arteries.
   b. vena cava.
   c. pulmonary arteries.
   d. ureters.
   e. renal veins.

10. During tubular secretion,
    a. blood cells and large proteins are removed from the blood.
    b. urea diffuses into the surrounding tissues, causing water to leave the filtrate.
    c. active transport adds ammonia, uric acid and excess H^+ to urine.
    d. salt is actively transported out of the filtrate.
    e. water, urea, glucose, amino acids, and ions are filtered from the blood.

Additional Study Help

Visit the ARIS (Assessment, Review, and Instruction System) site at aris.mhhe.com for quizzes, animations, and other study tools.

29   How the Animal Body Defends Itself

Key Concepts Outline

Skin and mucus membranes offer an efficient barrier to penetration by microbes. (pages 576-577)
- Surface defenses can be breached, and when they are, a second line of defense is used.

The second line of defense includes cellular and chemical attacks on the invading microbes. (pages 578-580)
- Cellular attacks are performed by macrophages, neutrophils and natural killer cells.
- Complement, inflammatory and temperature responses comprise the chemical attack.

Specific immunity is the third line of defense. (pages 581-587)
- Antibody production associated with B cells and the response by T cells provide specific immunity.
- The specific response is stimulated by chemicals produced by macrophages when they encounter cells without the proper MHC proteins.
- T cells are responsible for the cellular immune response; helper T cells activate cytotoxic T cells that actually carry out the cellular attack.
- B cells and the antibodies produced by plasma cells create the humoral immune response; memory cells provide long-term immunity.
- B and T cells that respond to infectious agents form clones of themselves that create a strong immune response.

Immune responses of vertebrates are similar, and analogous systems exist in invertebrates. (pages 588-589)

An immune response against a pathogen can be elicited by a vaccination of antigens that are similar or identical to those of the pathogen. (pages 590-591)
- The first vaccines developed were dead or disabled pathogens; current vaccinations are genetically engineered.

There are a variety of medical uses for antibodies. (page 592)

Autoimmune diseases and allergies are the result of overactive immune systems. (page 593)

HIV causes AIDS by disabling critical lymphocytes, resulting in the collapse of the immune system. (pages 594-595)

Key Terms Matching

1. _____ Epidermis	a. Mature in spleen and lymphoid tissue, produce protective proteins.
2. _____ Dermis	b. 20 different proteins that circulate in the blood in an inactive state.
3. _____ Stratum corneum	c. A response to bacterial infection creating unfavorable environment.
4. _____ Macrophage	d. Five classes of immunoglobulins.
5. _____ Neutrophils	e. Shed and replaced constantly.
6. _____ Complement system	f. Caused by histamines and prostaglandins.
7. _____ Inflammatory response	g. Artificial stimulation of immune system.
8. _____ Fever	h. Middle layer of skin.
9. _____ T cells	i. Antigen presentation.
10. _____ B cells	j. An overactive immune response.
11. _____ Helper T cell	k. They kill everything in the area with "Chlorox."
12. _____ Interleukin-1	l. Cause of acquired immunodeficiency syndrome.
13. _____ Antibodies	m. Outermost layer of skin.
14. _____ Vaccination	n. Four types, mature in the thymus.
15. _____ Allergy	o. White blood cell with a large appetite.
16. _____ HIV	p. Released in response to viral infection to protect neighboring cells.

29.1   Skin: The First Line of Defense (Pages 576–577)

17. Label the cross section of human skin (a – f).

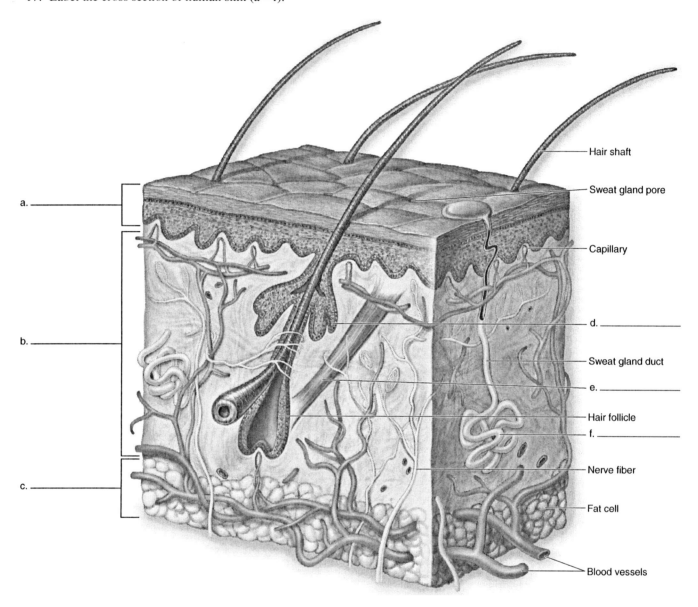

Hair shaft

Sweat gland pore

a. _____

Capillary

d. _____

Sweat gland duct

e. _____

b. _____

Hair follicle

f. _____

Nerve fiber

c. _____

Fat cell

Blood vessels

29.2   Cellular Counterattack: The Second Line of Defense (Pages 578–580)

18. Which of the following best describes the differences between macrophages and monocytes?
     a. Macrophages are phagocytic and monocytes release substances such as histamine into the bloodstream.
     b. Monocytes are phagocytic and macrophages release substances such as histamine into the bloodstream.
     c. Macrophages attack bacterial antigens exclusively and monocytes attack viral antigens exclusively.
     d. Macrophages are able to discriminate between bacterial and viral antigens but monocytes cannot.
     e. They are the same type of cell but macrophages remain at one site and monocytes move through the bloodstream.

19. Natural killer cells attack
    a. viruses only.
    b. bacteria only.
    c. parasitic protozoans only.
    d. cancerous cells only.
    e. infected body cells.

29.3 Specific Immunity: The Third Line of Defense (Page 581)

20. _____ Killer cell
21. _____ B cell
22. _____ Inducer T cell
23. _____ Mast cell

a. Initiates inflammation.
b. Mediates maturation of other cells in the same class.
c. Precursor to plasma cell.
d. One type only attacks antibody-covered cells.

29.4 Initiating the Immune Response (Page 582)

24. All of the body's cells have protein markers on their surfaces called _____ that enable the immune system to distinguish between self and nonself.

25. When _____ encounter cells without the proper _____ they initiate the immune response.

29.5 T Cells: The Cellular Response (Page 583)

26. When helper T cells are activated they secrete _____, which stimulates the proliferation of _____.
    a. antibodies, B cells
    b. antibodies, cytotoxic T cells
    c. interleukin-2, cytotoxic T cells
    d. interleukin-1, lymphokines
    e. lymphokines, interleukin-2

29.6 B Cells: The Humoral Response (Pages 584–585)

True (T) or False (F) Questions
If you believe the statement to be false, rewrite the statement as a true one.

27. IgA is found in bodily secretions such as breast milk and helps prevent pathogens from attaching to host tissues.
    Answer: _____ Restatement: _____
    _____

28. Memory T cells remain in the body for the rest of your life, making you immune to any antigen you have encountered.
    Answer: _____ Restatement: _____
    _____

29.7 Active Immunity Through Clonal Selection (Pages 586–587)

29. Joe Smith had the flu last year. This winter he "caught" the flu again. Why wasn't he immune the second time?
    a. Memory cells produced in response to viral infections only last for a short period of time, usually only several weeks.
    b. Joe obviously has a weak immune system.
    c. Many viruses, such as influenza viruses, are able to rapidly change their surface antigens, thereby avoiding the previously produced memory cells.
    d. Memory cells are only effective against bacteria and protozoans and are not able to recognize viral antigen.
    e. None of the above explain why Joe caught the flu this year.

29.8   Evolution of the Immune System (Pages 588–589)

30. Invertebrates possess proteins called _____ which may be the forerunners to antibodies.
31. The modern immune system was first seen in the _____.

29.9   Vaccination (Pages 590–591)

32. Research done on HIV has determined that mutations in the _____ render the virus nonvirulent by
_____.

33. Fortunately for Jenner and his patients, the cowpox virus was able to stimulate immunity to smallpox because
_____.

29.10   Antibodies in Medical Diagnosis (page 592)

34. A person with blood type AB would have
    a. A antigens.
    b. B antigens.
    c. both A and B antigens.
    d. no antigens.

35. Rh incompatibility generally occurs if
    a. the mother is Rh-positive and the fetus is Rh-negative.
    b. the mother is Rh-negative and the fetus is Rh-positive.
    c. both mother and fetus are Rh-negative.
    c. none of the above

29.11   Overactive Immune System (Page 593)

36. Examples of autoimmune diseases include _____, _____, _____, _____.

29.12   AIDS: Immune System Collapse (Pages 594–595)

True (T) or False (F) Questions
If you believe the statement to be false, rewrite the statement as a true one.

37. HIV is the highly contagious cause of acquired immunodeficiency syndrome.
    Answer: _____ Restatement: _____
    _____

38. HIV has an affinity for CD4$^+$ cells and has little effect on the number of CD8$^+$ cells in an infected individual.
    Answer: _____ Restatement: _____
    _____

Chapter Test

1. Cells in the stratum corneum live there for about
   a. 8 hours.   b. a day.   c. a week.   d. a month.   e. 2 to 3 months.

2. A membrane attack complex is
   a. a group of macrophages that are attacking an invading bacterium or virus.
   b. a group of activated complement molecules.
   c. necessary for mast cells to release histamines.
   d. found in viral infections only.
   e. formed when B cells secrete antibody molecules.

3. Complement is activated by
   a. exposure to proteins released by virus-infected cells.
   b. the lysis of neutrophils.
   c. encounter with bacterial or fungal cell walls.
   d. a decrease in vitamin C levels.
   e. both a and d.

4. The redness and swelling that are associated with inflammation are due to
   a. histamines.
   b. the action of neutrophils.
   c. chemical signals released from damaged cells.
   d. the production of interleukins by macrophages.
   e. waste products released by natural killer cells.

5. Macrophages
   a. phagocytize invading microbes in the body.
   b. produce interleukins.
   c. identify foreign cells in cooperation with B and T cells.
   d. examine the major histocompatibility proteins on the surface of the body's cells.
   e. all of the above.

6. Which of the following is the best description of the relationship between helper T cells and interleukin-1?
   a. Helper T cells secrete interleukin-1 when they encounter a virus-infected cell.
   b. Interleukin-1 is a surface protein found on helper T cells.
   c. Interleukin-1 is secreted by natural killer cells and it activates helper T cells.
   d. Macrophages secrete interleukin-1, which then activates helper T cells.
   e. Helper T cells secrete interleukin-1, which then activates macrophages.

7. Killer T cells help to eliminate infected body cells and cancer cells by
   a. secreting antibodies to mark the cells for destruction.
   b. phagocytosis.
   c. fusing with them and secreting histamines that cause them to lyse.
   d. puncturing the surface of the cell.
   e. both a and d.

8. The production of antibodies is initiated by the
   a. release of interleukin-1.
   b. release of interleukin-2.
   c. release of histamines.
   d. development of a fever.
   e. killer T cells.

9. Mutations in the *nef* gene
   a. render B cells ineffective.
   b. result in increased allergies.
   c. inhibit the release of histamines.
   d. render HIV nonvirulent.
   e. affect the ability of macrophages to release interleukin-1.

10. The symptoms of lupus are caused by
    a. destruction of the myelin on motor nerves by the immune system.
    b. an attack on the tissues of the joint by the immune system.
    c. an attack on the connective tissue and kidneys by the immune system.
    d. an attack on the thyroid by the immune system.
    e. destruction of pancreas cells by the immune system.

11. Multiple sclerosis is caused by
    a. destruction of the myelin on motor nerves by the immune system.
    b. an attack on the tissues of the joint by the immune system.
    c. an attack on the connective tissue and kidneys by the immune system.
    d. an attack on the thyroid by the immune system.
    e. destruction of pancreas cells by the immune system.

12. Graves disease is caused by
    a. destruction of the myelin on motor nerves by the immune system.
    b. an attack on the tissues of the joint by the immune system.
    c. an attack on the connective tissue and kidneys by the immune system.
    d. an attack on the thyroid by the immune system.
    e. destruction of pancreas cells by the immune system.

13. HIV is transmitted by
    a. unprotected sex.
    b. protected sex.
    c. exposure to contaminated needles.
    d. all of the above.
    e. both a and c.

Matching

14. _____ Helper T cells
15. _____ Mast cells
16. _____ Macrophages
17. _____ Natural killer cells
18. _____ Neutrophils
19. _____ B cells

a. These cells are efficient at detecting cells that are infected by viruses.
b. These cells help coordinate the activities of other immune cells.
c. These cells release substances that are like household bleach to kill bacteria.
d. A large number of these hungry cells are found in the spleen.
e. Receptor molecules called antibodies are produced by these cells.
f. These cells release histamines and other substances in allergic reactions.

Additional Study Help

Visit the ARIS (Assessment, Review, and Instruction System) site at aris.mhhe.com for quizzes, animations, and other study tools.

30   The Nervous System

Key Concepts Outline

During the evolution of the nervous system, associative activity localized in a brain has increased. (pages 600-601)
Neurons are specialized to conduct impulses resulting from ion movements across the neuron's plasma membrane. (pages 602-607)
- Na^+ and K^+ cross the membrane through protein channels that open or close in response to stimuli.
- Impulses cross synapses via neurotransmitters; addictive drugs affect synapses.

The central nervous system is composed of the brain and spinal cord. (pages 608-613)
- Forebrains became more prominent as vertebrates evolved.
- The cerebral cortex performs association activities; other parts of the brain process information, regulate body functions and coordinate muscle movements.
- A backbone in vertebrates protects the spinal cord.

Muscle contractions are consciously controlled by the somatic or voluntary nervous system; the autonomic nervous system sends commands to muscles and glands that are not controlled by conscious thought. (pages 614-615)
- The sympathetic and parasympathetic subdivisions of the autonomic nervous system are active during times of stress or rest and digestion, respectively.

A variety of sensory receptors initiate nerve impulses that transmit information about internal conditions essential to maintaining homeostasis and the external environment. (pages 616-626)

Key Terms Matching

1. _____ Central nervous system		a. Thin, gray, outer layer of the brain.
2. _____ Motor nerves		b. Neurons pump sodium out to maintain this.
3. _____ Sensory nerves		c. Sensory receptors initiate nerve impulses by opening or closing.
4. _____ Resting potential		d. Relays commands to smooth muscles.
5. _____ Voltage-gated channels		e. Open and close in response to changes in membrane potential.
6. _____ Neurotransmitters		f. Controls body temperature and heartbeat.
7. _____ Integration		g. Controls "fight or flight" response.
8. _____ Neuromodulators		h. Light-detecting complex.
9. _____ Cerebral cortex		i. Produces a rapid motor response to a stimulus.
10. _____ Hypothalamus		j. Canceling or reinforcing of excitatory and inhibitory effects.
11. _____ Autonomic nervous system		k. The brain and the spinal cord.
12. _____ Reflex		l. Detect greys even in low light.
13. _____ Sympathetic nervous system		m. Carry impulses to the CNS.
14. _____ Parasympathetic nervous system		n. Carry messages across synapses.
15. _____ Stimulus-gated channel		o. The back surface of the eye.
16. _____ Retina		p. Prolong transmission of signal across synapse.
17. _____ Rods		q. Conserves energy by slowing the heartbeat.
18. _____ Rhodopsin		r. Carry impulses to muscles and glands.

30.1   Evolution of the Animal Nervous System (Pages 600–601)

19. The nervous system consists of _____ and _____.
20. Through evolution more and more associative activity of the nervous system occurred and localized in the _____.

30.2 Neurons Generate Nerve Impulses (Pages 602-603)

21. Neurons receive nutritional support from
    a. the nodes of Ranvier.
    b. neuroglial cells.
    c. the array of dendrites which surround them.
    d. axons.
    e. oligodendrocytes.

22. The myelin sheath is produced by
    a. Schwann cells.   b. synaptic stimulation.   c. neuroglial cells.   d. myelin cells.   e. none of the above.

True (T) or False (F) Questions
If you believe the statement to be false, rewrite the statement as a true one.

23. Protein channels in the plasma membrane of neurons through which ions cross help to regulate nerve impulses.
    Answer: _____ Restatement: _____
    _____

24. Sodium ions are being moved out of a neuron and potassium ions into a neuron when it is at rest.
    Answer: _____ Restatement: _____
    _____

30.3 The Synapse (Pages 604-605)

25. Axons _____ when nerve impulses reach their ends.
    a. die
    b. produce and release serotonin
    c. become excited
    d. attach themselves to the dendrites of a neighboring neuron
    e. release a neurotransmitter

30.4 Addictive Drugs Act on Chemical Synapses (Pages 606-607)

26. Prozac helps battle depression by _____.
27. Drugs such as nicotine and cocaine act as powerful _____.

30.5 Evolution of the Vertebrate Brain (Pages 608-609)

28. _____ Midbrain   a. Associative activity – "end brain."
29. _____ Thalamus   b. Integration and relay of sensory information.
30 _____ Hypothalamus   c. Made up mostly of the optic lobes.
31. _____ Cerebrum   d. Controls secretions of the pituitary gland.

30.6 How the Brain Works (Pages 610-612)

32. Information is transmitted from the cortex to the rest of the brain by
    a. the secretion of serotonin into nerve synapses.
    b. creating action potentials by pumping sodium ions out of neurons.
    c. a solid white region of myelinated nerve fibers.
    d. the hypothalamus.
    e. the temporal lobe.

33. Which part of the brain enables birds to land safely on a branch moving in the wind?
    a. the thalamus   b. the brain stem   c. the limbic system   d. the cerebellum   e. the pons

30.7 The Spinal Cord (Page 613)

34. In humans and other vertebrates messages from the brain and the body move _____.
35. The outer edges of the spinal cord are white due to a covering _____.

30.8 Voluntary and Autonomic Nervous Systems (Pages 614-615)

True (T) or False (F) Questions
If you believe the statement to be false, rewrite the statement as a true one.

36. After a meal of fried chicken the voluntary nervous system would play a role in stimulating the gallbladder to release bile.
    Answer: _____ Restatement: _____
    _____

37. Reflexes cause the body to respond often before the cerebrum is aware danger exists.
    Answer: _____ Restatement: _____
    _____

38. The _____ system is responsible for bodily responses designed to conserve energy – such as slowing down the rate of the heart.
    a. voluntary   b. autonomic   c. parasympathetic   d. sympathetic   e. both a and c

30.9 Sensory Perception (Pages 616-617)

39. Sense stimuli arising from within the body are detected by _____ while _____ sense stimuli from the environment.
40. _____ are specialized cells that detect many stimuli such as smells and changes in blood pressure.

True (T) or False (F) Questions
If you believe the statement to be false, rewrite the statement as a true one.

41. While blood pressure is sensed by neurons called baroreceptors, changes in temperature are detected by two types of nerve endings.
    Answer: _____ Restatement: _____
    _____

30.10 Sensing Gravity and Motion (Page 618)

42. Semicircular canals allow an organism to sense
    a. smells.   b. balance.   c. changes in blood pressure.   d. temperature.   e. motion.

43. Movement of fluid inside the inner ear allows an organism to sense motion when
    a. the fluid changes temperature relative to the external environment.
    b. cilia on hair cells are displaced.
    c. the pressure inside the ear changes.
    d. Both a and c are true.
    e. None of the above are true.

30.11 Sensing Chemicals: Taste and Smell (Page 619)

44. Taste buds on the tongue are sensitive to four tastes. They are _____, _____, _____ and
    _____.

30.12   Sensing Sounds: Hearing (Pages 620-621)

45. Label the parts of the human ear (a – g).

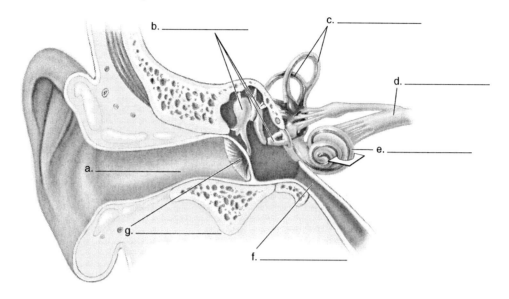

46. Bat sonar enables the organism to locate its "meal" by _____.
47. Fish rely on their _____ to swim in unison with their school.

30.13   Sensing Light: Vision (Pages 622–625)

48. _____ Cornea   a. Densely packed area with 3 million cones.
49. _____ Iris   b. Transparent protective covering of the eye.
50. _____ Pupil   c. Light-sensitive area containing rods and cones.
51. _____ Retina   d. Controls amount of light entering the eye.
52. _____ Fovea   e. Transparent zone in the middle of the iris.

53. Label the parts of the human eye (a – g).

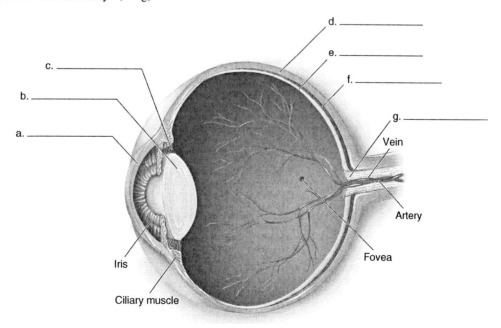

30.14 Other Types of Sensory Reception (Page 626)

True (T) or False (F) Questions
If you believe the statement to be false, rewrite the statement as a true one.

54. A pit organ is a cluster of photoreceptors which allow pit vipers to detect the location of their prey.
    Answer: _____ Restatement: _____
    _____

Chapter Test

1. The autonomic nervous system
   a. controls the skeletal muscles.
   b. is active only when the body is in crisis.
   c. stimulates glands and controls the smooth muscles.
   d. controls all of the body's functions.
   e. none of the above.

2. A stroke is caused by
   a. blockage of the heart by clots.
   b. buildup of cholesterol in blood vessels.
   c. blockage of blood vessels in the brain by clots.
   d. insufficient thyroid hormone.
   e. lack of oxygen.

3. The reticular formation
   a. is only present in the brains of epileptics.
   b. is a network of cells involved in arousal states.
   c. is sometimes called the medulla oblongata.
   d. is responsible for the control of smooth muscles.
   e. allows birds to maintain their balance as they fly.

4. The outer edges of the spinal cord are white because
   a. no red blood cells are present in the area.
   b. it is covered in a thick layer of connective tissue.
   c. the axons within it are covered in myelin.
   d. a layer of fat surrounds the structure.
   e. both a and c.

5. Interneurons are
   a. connecting neurons that are between the sensory and motor neurons.
   b. the spaces between dendrites and other neurons.
   c. stimulated by the pituitary gland.
   d. found on the epidermis.
   e. none of the above.

6. Interoreceptors
   a. regulate activity between adjacent neurons.
   b. detect information about the body's internal condition.
   c. work with neurotransmitters to regulate the speed of nerve impulses.
   d. open and close ion channels on sensory neurons.
   e. work only in the central nervous system.

7. A person with defective otolith sensory receptors
   a. has a difficult time maintaining balance.
   b. is deaf.
   c. cannot detect external temperature changes.
   d. has a faulty sense of smell.
   e. has a limited imagination.

8. Parallax is
   a. the synchronized functioning of neurons.
   b. a slight displacement of images that plays a role in distance perception.
   c. the path that light takes through the eye.
   d. only important in land animals.
   e. a defect caused by eyes that are oblong in shape.

Matching—Part 1

9. _____ Spinal cord   a. Balance, posture, and muscular coordination are controlled here.
10. _____ Cerebrum   b. Body temperature and blood pressure are controlled here.
11. _____ Hypothalamus   c. It is sometimes called the brain stem.
12. _____ Cerebellum   d. This structure consists of sensory and motor nerve tracts.
13. _____ Medulla oblongata   e. The "center for higher thought."

Matching—Part 2

14. _____ Neuroglial cells   a. Nerve impulses along the axon are facilitated by them.
15. _____ Myelin sheath   b. These extend from one end of a cell body.
16. _____ Nodes of Ranvier   c. They make up more than half the volume of the human
17. _____ Dendrites   nervous system.
18. _____ Schwann cells   d. A layer of insulation surrounding the axon.
                                          e. Gaps along the length of an axon.

Additional Study Help

Visit the ARIS (Assessment, Review, and Instruction System) site at aris.mhhe.com for quizzes, animations, and other study tools.

31 Chemical Signaling Within the Animal Body

Key Concepts Outline

Many of the hormone-producing endocrine glands are under the direct control of the central nervous system. (pages 630-631)

- The hypothalamus produces releasing hormones, which cause the pituitary to release particular hormones that circulate via the bloodstream to their target tissues.
- Only cells with an appropriate receptor will respond to a specific hormone.

Steroid and peptide hormones affect their target cells differently. (pages 632-633)

- Steroid hormones are lipid-soluble and interact with receptors located in the cytoplasm or nucleus.
- Peptide hormones interact with receptors located on the cell surface and initiate cAMP, a second messenger.

Hypothalamic releasing and inhibiting hormones control the production of several hormones by the anterior pituitary gland; the posterior pituitary gland stores and releases two hormones produced by the hypothalamus. (pages 634-636)

- ADH affects water retention by the kidneys and oxytocin stimulates milk release and uterine contractions; both are released by the posterior pituitary gland.
- Hormones made by the anterior pituitary gland affect a number of other endocrine organs.
- The production of many hormones is controlled by negative feedback.

Various other endocrine glands contribute to maintaining homeostasis and regulating reproduction. (pages 637-642)

- Specific pancreatic cells produce insulin, which decreases blood glucose, and glucagon, which increases blood glucose.
- The thyroid secretes thyroxine, which affects the metabolic rate and growth; blood calcium is regulated by the thyroid's calcitonin, and parathyroid hormone is made by the parathyroids.
- The adrenal medulla secretes epinephrine; the adrenal cortex secretes cortisol, which regulates glucose balance, and aldosterone, which regulates Na^+ and K^+ balance.
- The ovaries and testes produce the sex steroid hormones, including estrogens, progesterone, and testosterone.
- Blood oxygen is regulated by erythropoietin that is made by the kidneys.
- The pineal gland in the brain secretes the hormone melanin in response to a hypothalamic biological clock.

Key Terms Matching

1. _____ Endocrine gland	a. Bind to receptors embedded in the cell membrane.
2. _____ Neuroendocrine system	b. Makes up the two major routes to the central nervous system.
3. _____ Steroid hormones	c. Also called ecdysone.
4. _____ Anabolic steroids	d. Produces nine major hormones.
5. _____ Peptide hormones	e. They are completely enclosed in tissue.
6. _____ Second messengers	f. Amplify the original signal.
7. _____ Pituitary gland	g. Synthetic compounds resembling testosterone.
8. _____ Negative feedback inhibition	h. Secretes melatonin.
9. _____ Diabetes mellitus	i. Inability to take glucose up from the blood.
10. _____ Pineal gland	j. Lipid-soluble molecules.
11. _____ Molting hormone	k. One hormone inhibiting secretion of another hormone.

31.1 Hormones (Pages 630–631)

12. Chemical communication in the body involves two elements: a(n) _____ and a(n) _____.

13. Hormones can act on _____ or be transported _____.

31.2   How Hormones Target Cells (Pages 632–633)

14. Steroid hormones are manufactured from _____, a molecule with _____ carbon rings.
15. Steroid hormones bind to receptors located _____.

16. In response to binding of a peptide hormone to its receptor,
    a. a second messenger is activated.
    b. the adrenal grand is stimulated to secrete adrenaline.
    c. serotonin is released from the cell.
    d. the organism's temperature increases.
    e. the organism's heartbeat slows.

17. In response to binding with insulin, its receptor
    a. changes shape.
    b. loses the ability to sense glucose concentration.
    c. gains the ability to sense glucose concentration.
    d. binds with additional peptide receptors.
    e. None of the above

31.3   The Hypothalamus and the Pituitary (Pages 634–636)

18. _____ Melanocyte-stimulating hormone   a. Causes color changes in the skin of reptiles.
19. _____ Luteinizing hormone   b. Stimulates muscle growth.
20. _____ Thyroid-stimulating hormone   c. Stimulates production of testosterone by male gonads.
21. _____ Somatotropin   d. Eventually stimulates oxidative respiration.

31.4   The Pancreas (Page 637)

True (T) or False (F) Questions
If you believe the statement to be false, rewrite the statement as a true one.

22. 90% of diabetics suffer from type I diabetes.
    Answer: _____ Restatement: _____
    _____

23. In type II diabetes the number of insulin receptors is high.
    Answer: _____ Restatement: _____
    _____

31.5   The Thyroid, Parathyroid and Adrenal Glands (Pages 638–639)

24. Epinephrine is produced by the _____ while cortisol is a product of the _____.
    a. kidney, adrenal gland   d. adrenal gland, kidney
    b. adrenal medulla, adrenal cortex   e. thyroid, adrenal gland
    c. thyroid, parathyroid

25. Which of the following has an effect on calcium regulation?
    a. Calcitonin   d. All of the above
    b. Parathyroid hormone   e. Only a and b
    c. Parathyroid gland

26. An entomologist is conducting research on a species of butterfly native to New York state. She wants to determine how to stimulate the butterfly to more quickly enter the larval stage so she can determine how many calories are consumed at that point in the insect's life cycle. She should be able to achieve her goal by administering
    a. molting hormone.   d. epinephrine.
    b. melatonin.   e. a combination of molting hormone and epinephrine.
    c. juvenile hormone.

Chapter Test

1. Hormones
   a. are relatively unstable and work only in the area adjacent to the gland that produced them.
   b. are stable, long-lasting chemicals released from glands.
   c. typically act at a site that is distant from where it was produced.
   d. are all lipid-soluble.
   e. both b and c.

2. Glands that are enclosed by tissue are called
   a. internal glands.   d. pseudoglands.
   b. protected glands.   e. none of the above.
   c. endocrine glands.

3. The central nervous system issues commands to the body's organs by means of
   a. the motor nervous system.   d. both a and b.
   b the endocrine system.   e. none of the above.
   c. electrical impulses only.

4. How can a target cell recognize a particular hormone and not respond to other hormones?
   a. Target cells respond to whatever hormone is present in the largest concentration.
   b. The target cells are always located close to the source of the hormone, making it easy to respond to that hormone.
   c. Protein receptors located on the surface of the target cell or in the cytoplasm match the hormone.
   d. Carbohydrate tags on the surface of the target cell match the hormone.
   e. The hormone is only able to enter membrane channels on the correct target cell.

5. An example of a steroid hormone that influences secondary sexual characteristics is
   a. cortisone.   d. cyclic AMP.
   b. prolactin.   e. testosterone.
   c. follicle-stimulating hormone.

6. The steroid hormones bind to
   a. protein receptors in the cytoplasm of the target cell.
   b. protein receptors on the surface of the target cell.
   c. carbohydrate receptors in the cytoplasm of the target cell.
   d. carbohydrate receptors on the surface of the target cells.
   e. both a and c.

7. Low blood calcium levels cause the parathyroid gland to
   a. increase in size to form a goiter.   d. produce additional quantities of PTH.
   b. inhibit the release of calcium from the bones.   e. none of the above.
   c. release insulin.

8. Individuals with diabetes mellitus
   a. are unable to take up glucose from the blood.
   b. are unable to break down glucose.
   c. require additional insulin to break down glucose.
   d. are twice as likely as nondiabetics to have a heart attack.
   e. cannot consume glucose in their diet.

9. Individuals with type I diabetes
   a. have an autoimmune disease in which the islets of Langerhans are attacked.
   b. are allergic to insulin.
   c. have an abnormally low number of insulin receptors on target cells.
   d. often die before five years of age.
   e. both a and b.

10. Individuals with type II diabetes
    a. have an autoimmune disease in which the islets of Langerhans are attacked.
    b. are allergic to insulin.
    c. have an abnormally low number of insulin receptors on target cells.
    d. often die before five years of age.
    e. both a and b.

Matching

11.	_____ Oxytocin	a. Stimulation of muscle and bone growth.
12.	_____ ACTH	b. Amplification of an original hormone signal.
13.	_____ Thyroxine	c. Water retention by the kidneys.
14.	_____ Parathyroid hormone	d. The initiation of milk release in mothers.
15.	_____ Somatotropin	e. Regulation of blood calcium levels.
16.	_____ Melanocyte-stimulating hormone	f. Contains iodine.
17.	_____ Vasopressin	g. Stimulation of the adrenal gland to produce steroid hormones.
18.	_____ Second messengers	h. Stimulation of color changes in reptiles.

Additional Study Help

Visit the ARIS (Assessment, Review, and Instruction System) site at aris.mhhe.com for quizzes, animations, and other study tools.

32 Reproduction and Development

Key Concepts Outline

Animals commonly reproduce sexually, but many reproduce by a variety of asexual means. (pages 646-647)
- Different individuals of a species generally contribute the gametes that fuse during sexual reproduction.

During sexual reproduction fertilization may be external or internal, and the development that follows may be external or may occur in a shelled egg or within the female. (pages 648-651)
- External fertilization and development is characteristic of most fishes and amphibians.
- The embryos of birds and most reptiles develop in shelled eggs.
- Most mammals give birth to live offspring that develop inside the female.

Male gametes (sperm) are produced in the testes in very large numbers. (pages 652-653)
- Sperm mature in the epididymis and travel through the vas deferens to the urethra.

Female gametes (eggs) develop from oocytes located in the ovaries. (pages 654-655)
- An egg travels down a fallopian tube following ovulation.
- If fertilization occurs in the fallopian tube, the zygote will implant in the uterine wall.

The reproductive cycle is regulated by several hormones produced by the pituitary gland and ovary. (pages 656-657)
- An average human menstrual cycle is 28 days in length; the cycle has two distinct phases, follicular and luteal.

Embryonic development occurs in three phases: cleavage, gastrulation and neurulation; the organs begin developing in the fourth week of human pregnancy. (pages 658-663)
- Growth is primarily what occurs after the third month of human pregnancy.

Many methods of birth control exist, some more effective than others; condom use can also reduce the spread of many sexually transmitted diseases. (pages 665-666)

Key Terms Matching

1. _____	External fertilization	a. Also known as oviducts.
2. _____	Oviparity	b. Introduction of male gametes into female reproductive tract.
3. _____	Viviparity	c. These are stimulated by FSH.
4. _____	Internal fertilization	d. Young develop within mother and obtain nutrients from her blood.
5. _____	Estrus	e. Occurs 10 – 11 days after fertilization, a cell migration.
6. _____	Sperm	f. Avoid having intercourse just before and just after ovulation.
7. _____	Oocytes	g. Occurs if fertilization does not quickly follow ovulation.
8. _____	Fallopian tubes	h. Most of these never receive the "go" signal.
9. _____	Follicular phase	i. Most of the bony fishes reproduce using this method, releasing sperm in water.
10. _____	Follicle	j. Stage following morula.
11. _____	Luteal phase	k. Embryo limbs assume adult shape.
12. _____	Menstruation	l. A coil placed in the uterus to prevent conception.
13. _____	Blastula	m. Male gametes.
14. _____	Gastrulation	n. Eggs are fertilized internally and deposited to develop externally.
15. _____	Neurulation	o. Formation of notochord and hollow dorsal nerve cord.
16. _____	Morphogenesis	p. The period of sexual receptivity.
17. _____	Rhythm method	q. The hormone that causes ovulation is released at this time.
18. _____	IUD	r. When the egg develops within the ovary.

32.1   Asexual and Sexual Reproduction (Pages 646-647)

19. The development of eggs without fertilization is called _____ and occurs in _____.
    a. hermaphroditism, mammals
    b. hermaphroditism, arthropods
    c. parthenogenesis, mammals
    d. parthenogenesis, arthropods
    e. spontaneous fertilization, amphibians

20. Protogyny and protandry seem to be controlled by:
    a. social pressures.
    b. the ability of individuals in the school of fishes to quickly change gender.
    c. the ability of the queen bee to store sperm for long periods of time.
    d. the ability to allow eggs to develop into drones without fertilization.
    e. none of the above

32.2   Evolution of Reproduction Among the Vertebrates (Pages 648-651)

21. A difference between ovoviviparity and viviparity is that in
    a. ovoviviparity fertilization of the egg is not necessary to produce another generation while fertilization is necessary in viviparity.
    b. ovoviviparity the embryo obtains nutrients from the egg yolk while in viviparity the embryo obtains nutrients from the mother's blood.
    c. ovoviviparity several offspring are produced while in viviparity only one offspring is produced.
    d. ovoviviparity the offspring obtains nutrients from the mother's blood while it is in the eggs while in viviparity the embyro receives nutrients from the yolk.
    e. none of the above.

22. Birds, most reptiles and _____ lay water-tight eggs.
    a. amphibians   b. bony fishes   c. monotremes   d. alligators   e. sharks

32.3   Males (Pages 652-653)

23. Label the parts of the sperm (a – d).

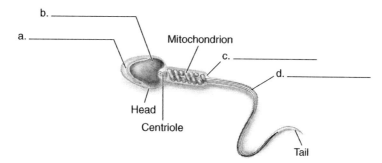

32.4   Females (Pages 654–655)

24. Label the parts of the human female reproductive system (a – e).

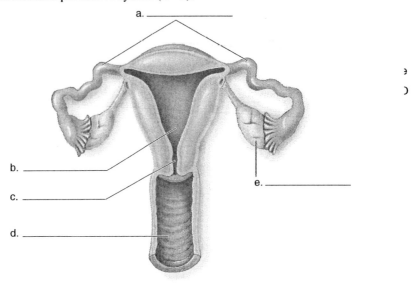

a. _____

b. _____

c. _____

d. _____

e. _____

32.5   Hormones Coordinate the Reproductive Cycle (Pages 656–657)

25. Following the release of follicle-stimulating hormone is the resumption of _____.
26. After childbirth the hormone _____ stimulates milk production while the hormone _____
    stimulates the release of milk.

32.6   Embryonic Development (Pages 658-659)

27. _____ Cleavage   a. Give rise to muscles, vertebrae, and connective tissues.
28. _____ Chorion   b. Interacts with placental tissues to form placenta.
29. _____ Blastocoel   c. During this stage there is no overall change in the size of the embryo.
30. _____ Notochord   d. A fluid-filled cavity.
31. _____ Somites   e. A flexible rod characteristic of vertebrates.

32.7   Fetal Development (Pages 660–663)

32. _____ is fine body hair that covers the embryo during the fifth month of development.
33. One of the leading causes of birth defects is caused by alcohol use by the mother and is called _____.

32.8   Contraception and Sexually Transmitted Diseases (Pages 665–666)

34. _____ Rhythm method   a. Prevent production of FSH and LH.
35. _____ Birth-control pills   b. Helps prevent delivery of sperm to the uterus.
36. _____ Interuterine device   c. Prevents implantation of embryo within uterine wall.
37. _____ Condom   d. Avoid sex 2 days prior to and after ovulation.

Chapter Test

1. The scrotum is located between the legs of the male rather than inside the body because
   a. that is a safer location for the testes.
   b. the temperature is cooler there, allowing sperm to complete their development.
   c. the external location makes it easier to transport sperm to the penis during ejaculation.
   d. the higher temperatures that occur there allow the sperm to develop normally.
   e. both c and d.

2. Men with sperm counts of fewer than _____ sperm per milliliter are considered sterile.
   a. 600 million   b. 250 million   c. 100 million   d. 50 million   e. 20 million

3. Progesterone is produced
   a. only in postmenopausal women.
   b. by the ovaries.
   c. by the Fallopian tubes.
   d. by the corpus luteum.
   e. just after menstruation.

4. A blastomere is
   a. the first opening that occurs in a blastula.
   b. a fluid-filled cavity.
   c. a cell in a morula.
   d. found in the mesoderm of the gastrula.
   e. an invagination on the surface of the embryo.

5. A blastocoel is
   a. the first opening that occurs in a blastula.
   b. a fluid-filled cavity.
   c. a cell in a morula.
   d. found in the mesoderm of the gastrula.
   e. an invagination on the surface of the embryo.

6. The miniature limbs of the embryo assume their adult shapes
   a. at three weeks of development.
   b. during the second month of development.
   c. during the third month of development.
   d. during the second trimester.
   e. during the third trimester.

7. Which of the following is the most effective form of birth control?
   a. condom
   b. birth-control pill
   c. rhythm method
   d. vasectomy
   e. coitus interruptus

Matching

8. _____ Epididymis   a. Primarily used for storage of sperm.
9. _____ Seminiferous tubules   b. This has an acrosome.
10. _____ Testosterone   c. A sac that is about $3°C$ cooler than the rest of the body.
11. _____ Sperm   d. Many tightly coiled tubes.
12. _____ Vas deferens   e. Where the reproductive and urinary tracts meet.
13. _____ Scrotum   f. The sperm-producing organs.
14. _____ Urethra   g. The maturation of sperm occurs here.
15. _____ Testes   h. A steroid hormone.

Additional Study Help

Visit the ARIS (Assessment, Review, and Instruction System) site at aris.mhhe.com for quizzes, animations, and other study tools.

33   Plant Form and Function

Key Concepts Outline

A vascular plant is organized along a vertical axis. (page 670)
- The shoot is the part aboveground and the root is the part belowground.
- Growth occurs in actively dividing zones called meristems.

There are a variety of ground, dermal and vascular tissues in a plant. (pages 671-672)
- Vascular tissue includes xylem that transports water and phloem that transports carbohydrates.

The vegetative components of a plant include roots that absorb water and minerals from the soil, stems that are supportive and leaves that perform photosynthesis. (pages 674-679)
- Growth from the tip of the shoot or root is called primary growth; growth in girth is called secondary growth and it takes place in meristem called the vascular or cork cambium.
- Strands of vascular tissue in stems occur as a cylinder toward the edge of the stem in dicots and scattered throughout the stem in monocots.
- Most leaves consist of a flattened blade and stalk called a petiole. Leaf arrangements on the stem exhibit considerable variation.

Transpiration draws water upwards from the roots to the leaves. (pages 680-682)
- Water molecules' tendency to form hydrogen bonds with each other forms an unbroken column of water.

Translocation is the means by which carbohydrates are moved throughout the plant. (page 683)
- Sugars move from sources to sink by a passive osmotic process.

Among the nutrients plants require are nitrogen, phosphorus and potassium. (page 684)

Key Terms Matching

1. _____ Shoot	a.	Responsible for most water uptake.
2. _____ Meristems	b.	Aboveground portion of a plant.
3. _____ Primary growth	c.	Control uptake of carbon dioxide.
4. _____ Secondary growth	d.	Major water loss occurs this way.
5. _____ Stomata	e.	Secondary xylem growth.
6. _____ Vascular cambium	f.	Growth zones of unspecialized cells.
7. _____ Mesophyll	g.	Water bonding to another substance.
8. _____ Parenchyma	h.	Most common plant cell.
9. _____ Wood	i.	Growth at tips.
10. _____ Casparian strip	j.	Activity of apical meristems.
11. _____ Root hairs	k.	Located between bark and main stem.
12. _____ Cohesion	l.	Surrounds endodermis cells.
13. _____ Transpiration	m.	"middle leaf."

33.1   Organization of a Vascular Plant (Page 670)

14. The two kinds of lateral meristems are _____ and _____.
15. The only function of meristems is _____.

33.2   Plant Tissue Types (Pages 671-674)

True (T) or False (F) Questions
If you believe the statement to be false, rewrite the statement as a true one.

16. Vascular tissue is made of xylem, which conducts water, and phloem, which conducts nutrients.
    Answer: _____ Restatement: _____

_____

17. Sclerenchyma cells do not contain cytoplasm when mature.
    Answer: _____ Restatement: _____

_____

33.3   Roots (Pages 674-675)

18. Label the parts of the root (a – f).

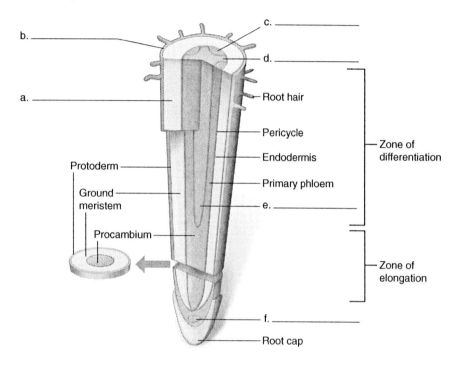

33.4   Stems (Pages 676-677)

19. As they begin to grow, young leaves appear around the _____.
20. Stems grow in two places: _____ and _____.

33.5   Leaves (Pages 678-679)

21. Label the parts of the leaf (a – g).

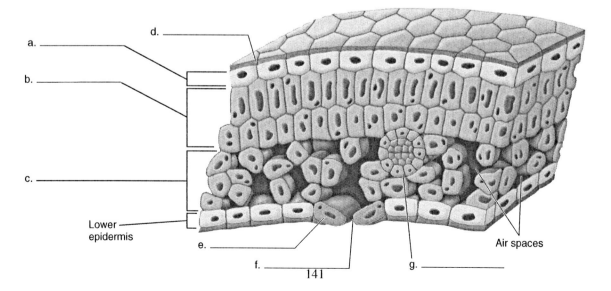

33.6   Water Movement (Pages 680-682)

22. Of all the water taken in by plants _____ % is lost by transpiration.
    a. 10   b. 25   c. 50   d. 75   e. 90

23. Transpiration is regulated by
    a. the number of hours of daylight in the day.   d. the thickness of the cuticle.
    b. opening and closing of stomata.   e. none of the above.
    c. the ability of water molecules to adhere to plant tissue.

33.7   Carbohydrate Transport (Page 683)

24. Carbohydrates are moved throughout the plant because of _____ due to osmosis.
25. When aphids attack plants they do so to obtain _____.

33.8   Essential Plant Nutrients (Page 684)

26. Turgor pressure in plants is regulated by _____.

Chapter Test

1. Which of the following cell types has no living cytoplasm at maturity?
   a. tracheids   b. sieve-tube members   c. companion cells   d collenchyma   e. meristems

2. Which of the following cell types lacks nuclei at maturity?
   a. tracheids   b. sieve-tube members   c. companion cells   d collenchyma   e. meristems

3. Which of the following cell types retains the ability to differentiate into another cell type if the need should arise in a plant?
   a. tracheids   b. sieve-tube members   c. companion cells   d collenchyma   e. meristems

4. Pits allow water to move through
   a. vessel elements.   b. tracheids.   c. companion cells.   d. sieve-tube members.   e. both a and d.

5. Leaf growth is due primarily to _____ meristems.
   a. apical   b. marginal   c. lateral   d. vertical   e. horizontal

6. Large intercellular spaces that function in gas exchange are found between cells of
   a. meristem.   b. stipules.   c. spongy mesophyll.   d. ground tissue.   e. cortex.

7. Periderm is
   a. cork cells.   d. all of the above.
   b. a dense layer of parenchyma.   e. none of the above.
   c. the cork cambium.

8. While collecting firewood to heat their mountain cabin, a family finds a dead juniper tree that has fallen over. After cutting the trunk into lengths, they notice the annual rings and count them. There are 34 rings. This means that the tree is
   a. 34 years old.   d. 17 years old.
   b. approximately 34 years old.   e. The growth rings do not really reflect the age of the tree.
   c. 340 years old.

9. A plant with damaged root hairs would
    a. be unable to absorb water.
    b. be unable to absorb minerals.
    c. rely on the root for water absorption.
    d. both a and b.
    e. both b and c.

10. Tensile strength can be increased by
    a. increasing the diameter of the tube of water.
    b. decreasing the diameter of the tube of water.
    c. adding an emulsifier to the water.
    d. increasing the temperature of the water.
    e. decreasing the temperature of the water.

11. While preparing for a dinner party at your house, you cut up some celery to eat with dip. You get busy and forget to put the celery in the refrigerator for a while and it becomes slightly soft. You place the celery in the refrigerator in some water and an hour later it is once again crispy. Why?
    a. The turgor pressure has been restored.
    b. The cold temperature made the celery crispy.
    c. Water has adhered to the celery, making it crispy.
    d. The cold temperature has activated enzymes that play a role in rigidity.
    e. None of the above.

Matching

12. _____ Fibers
13. _____ Tracheids
14. _____ Trichomes
15. _____ Sclereids
16. _____ Petiole
17. _____ Companion cells
18. _____ Parenchyma
19. _____ Guard cells
20. _____ Collenchyma

a. The least specialized and the most common of plant cell types.
b. A slender stalk that is attached to the leaf.
c. They are alive at maturity and form strands under the epidermis of stems.
d. A type of parenchyma that is associated with sieve-tube members.
e. Long, slender sclerenchyma cells that form strands.
f. They have pits and help make up xylem.
g. They are variable in shape but are often branched; may be called stone cells.
h. Paired cells that make up the stomata.
i. These outgrowths of the epidermis may make a leaf look "fuzzy."

Additional Study Help

Visit the ARIS (Assessment, Review, and Instruction System) site at aris.mhhe.com for quizzes, animations, and other study tools.

34   Plant Reproduction and Growth

Key Concepts Outline

Reproduction in angiosperms involves asexual and sexual reproduction. (pages 688-690)
- Asexual reproduction means of reproduction are: runners, rhizomes, suckers, and adventitious plantlets.
- Sexual reproduction occurs when pollinators transfer pollen to the flower's stigma; sperm travel to an egg via a pollen tube.
- Double fertilization leads to the formation of an embryo and endosperm.

Seeds contain an embryo and stored food encased in a protective seed coat. (page 691)
- Germination of the seed occurs when it is likely that the plant will survive.

Fruits disperse seeds away from the parent plant. (page 692)
- Common methods of dispersal include wind, water, attachment to animals or consumption by animals.

Water triggers germination of a seed's embryo. (page 693)
- The emergence of roots is the first stage of germination; further development is dependent on whether the plant is a monocot or dicot.

Auxin is produced at the tips of shoots and is transported downward. Auxin also migrates away from light and promotes the elongation of plant cells on the dark side, causing stems to bend in the direction of light. (pages 694-699)
- Cytokinins are necessary for mitosis and cell division in plants. They promote the growth of lateral buds and inhibit the formation of lateral roots.
- Gibberellins, along with auxin, play a major role in stem elongation in most plants.
- Ethylene influences leaf abscission and is widely used to hasten fruit ripening.
- Abscisic acid promotes the formation of winter buds and plays a key role in closing stomata.

The period of dark is actually the critical factor in initiating flowering in many plants. (pages 700-701)
- A molecule known as phytochrome occurs in two interconvertible forms, and plays a role in determining the flowering response.
- Dormancy is a plant adaptation that carries a plant through unfavorable seasons.
- A phototropism is a response to light, a gravitropism is a response to gravity and a thigmotropism is a response to contact.

Key Terms Matching

1. _____	Calyx	a. Unique feature in the life cycle of angiosperms.
2. _____	Corolla	b. Regulates cell growth in plants.
3. _____	Androecium	c. Plants growing so they bend toward light.
4. _____	Gynoecium	d. Collective term for the petals.
5. _____	Double fertilization	e. How organisms measure seasonal changes in day length.
6. _____	Phototropism	f. It means "female household."
7. _____	Auxin	g. Light receptors.
8. _____	Photoperiodism	h. How plants respond to touch.
9. _____	Phytochromes	i. This causes stems to grow upwards.
10. _____	Gravitropism	j. It means "male household."
11. _____	Thigmotropism	k. All of the sepals.

34.1   Angiosperm Reproduction (Pages 688-690)

12. Pollen grains develop from _____ formed in the four pollen sacs.
13. Pollen grains make their way to the ovary by means of the _____.

34.2   Seeds (Page 691)

True (T) or False (F) Questions
If you believe the statement to be false, rewrite the statement as a true one.

14. In order for pollination to occur, the pollen must be placed on the carpel of the flower.
    Answer: _____ Restatement: _____

    _____

15. A seed is a dormant embryo waiting for water to stimulate germination.
    Answer: _____ Restatement: _____

    _____

34.3   Fruit (Page 692)

True (T) or False (F) Questions
If you believe the statement to be false, rewrite the statement as a true one.

16. Seeds are not damaged by passage through the digestive system of animals.
    Answer: _____ Restatement: _____

    _____

34.4   Germination (Page 693)

17. Seeds require _____ and _____ for growth to begin.
18. After germination occurs the _____ perform the first photosynthesis.

34.5   Plant Hormones (Pages 694–695)

19. The five major hormones found in plants are:

    _____, _____, _____, _____, _____.

34.6   Auxin (Pages 696-697)

20. Auxin promotes growth in plants by
    a. increasing the plasticity of plant cell walls.   d. increasing the rate of photosynthesis.
    b. stimulating the uptake of water by seeds.   e. promoting root tip growth.
    c. causing elongation of stems.

21. Plants synthesize auxin from the amino acid
    a. cystine.   b. phenylalanine.   c. ornithine.   d. tryptophan.   e. lysine.

34.7   Other Plant Hormones (Pages 697-699)

22. Although more than _____ gibberellins have been identified only _____ of them stimulates shoot
    elongation.
23. Abscisic acid affects the closing of _____ by forcing potassium ions out of _____.

34.8   Photoperiodism and Dormancy (Pages 700-701)

24. To brighten her home in the winter, a woman decides to force some iris to bloom. To achieve this she should
    a. expose the plant to long periods of darkness.
    b. expose the plant to short periods of darkness.
    c. interrupt a long period of darkness with a flash of light.
    d. water it generously.
    e. both b and c

25. Many plants have seeds that are stimulated to germinate by
    a. exposure to fire.   d. exposure to ethylene.
    b. exposure to red light.   e. insect pollinators.
    c. auxin.

34.9   Tropisms (Page 701)

26. _____ are the responses of plants to external stimuli such as _____ , _____
    and _____ .

Chapter Test

1. In angiosperms, male and female reproductive structures usually develop
    a. in separate plants.   d. at different times so self-fertilization is avoided.
    b. on different flowers on the same plant.   e. at the same time so self-fertilization is ensured.
    c. in the same flower.

2. Angiosperms that rely on wind for pollination must
    a. produce copious amounts to ensure that pollination occurs.
    b. grow close together to ensure pollination.
    c. produce a foul odor to discourage insect pollinators.
    d. produce a foul odor to discourage bird pollinators.
    e. a, b and d are true.

3. Insects are encouraged to visit flowers by
    a. a food such as nectar.   d. a pleasant odor.
    b. flower colors that stimulate mating in the insects.   e. pheromones.
    c. the protection they provide from predators.

4. The embryo of a flowering plant is covered by the
    a. cuticle.   b. waxy layer.   c. integument.   d. cotyledon.   e. ground meristem.

5. In angiosperms, tissue differentiation in the embryo begins
    a. when the seed germinates.   d. almost immediately after fertilization.
    b. when the basal cell is stimulated by auxin.   e. after the cotyledons are fully grown.
    c. after the flower dies.

6. This plant hormone promotes stem elongation and enzyme production in developing seeds.
    a. auxin   b. cytokinin   c. ethylene   d. abscisic acid   e. gibberellins

7. This is produced by aging fruit.
    a. auxin   b. cytokinin   c. ethylene   d. abscisic acid   e. gibberellins

8. This hormone promotes lateral bud dormancy.
    a. auxin   b. cytokinin   c. ethylene   d. abscisic acid   e. gibberellins

9. This plant hormone promotes chloroplast development.
    a. auxin   b. cytokinin   c. ethylene   d. abscisic acid   e. gibberellins

10. This plant hormone inhibits the effects of other hormones.
    a. auxin   b. cytokinin   c. ethylene   d. abscisic acid   e. gibberellins

Matching

11. _____ Cytokinins   a. This hormone causes leaves to age and fall off of the plant.
12. _____ Abscisic acid b. This hormone would cause the stem to grow toward a source of light.
13. _____ Ethylene   c. This hormone is produced in the apical portion of stems.
14. _____ Gibberellins   d. Tomatoes are stimulated to ripen when exposed to this hormone.
15. _____ Auxins   e. Cell division is stimulated by this hormone.

Additional Study Help

Visit the ARIS (Assessment, Review, and Instruction System) site at aris.mhhe.com for quizzes, animations, and other study tools.

Answer Key

Chapter 1

Key Terms Matching
1. d   2. e   3. h   4. c   5. b   6. f   7. g
8. a

Review Questions
9. 6   10. Archaebacteria   11.   Plantae   12. d
13. c   14. e   15. Metabolism   16. DNA, genes
17. a   18. b   19. c   20. Tissues   21. a community,
the physical environment   22. b   23. c   24. a
25. Darwin, *On the Origin of Species*   26. deductive
27. inductive   28. b   29. c   30. a   31. a   32. e
33. b   34. scientific method   35. Supernatural and
religious   36. genome   37. chromosomes

Chapter Test
1. c   2. b   3. e   4. b   5. c   6. b   7. e   8. a   9. e
10. b   11. c   12. c   13. a   14. b

Chapter 2

Key Terms Matching
1. f   2. j   3. i   4. g   5. a   6. c   7. e   8. h   9. d
10. b

Review Questions
11. T   12. F Darwin's God produced changes in
species over time through evolution.   13. F Darwin
and Wallace were the first to propose natural selection
as the mechanism of evolution   14. a   15. c   16. e
17. c   18. d   19. a   20. e   21. b   22. d   23. c
24. b   25. a   26. b   27. a   28. c   29. b   30. d
31. b   32. c   33. b   34. interspecific   35. b
36. habitat, niche   37. character displacements   38. T
39. coevolution   40. mutualism   41. parasitism

Chapter Test
1. e   2. d   3. b   4. b   5. a   6. d   7. c   8. a   9. e
10. a   11. c   12. d   13. c

Chapter 3

Key Terms Matching
1. e   2. i   3. g   4. h   5. d   6. j   7. a   8. f   9. b
10. c

Review Questions
11. b   12. d   13. a   14. c   15. d   16. b   17. a
18. b   19. a   20. strong, directional   21. strong,
directional   22. polar   23. directional   24. attracted
25. cohesion, adhesion   26. hydrophobic   27. below,
above   28. H^+, OH^- 29. buffer

Chapter Test
1. d   2. d   3. c   4. b   5. a   6. b   7. e   8. a   9. d
10. c   11. a   12. a   13. b   14. d

Chapter 4

Key Terms Matching
1. j   2. g   3. i   4. f   5. d   6. c   7. h   8. e   9. b
10. a

Review Questions
11. dehydration synthesis   12. proteins   13.
monomers   14. Keratin, structural   15. lower the
energy required   16. polymer, amino acids   17.
quaternary   18. Chaperone   19. a five-carbon sugar, a
phosphate group, a nitrogenous base   20. thymine,
guanine   21. hydrogen bonds   22. a five carbon sugar,
a phosphate group   23. polysaccharides   24.
dehydration synthesis   25. C-H bonds   26. chitin
27. glycogen, starch   28. b   29. c   30. a

Chapter Test
1. c   2. d   3. a   4. e   5. e   6. d   7. a   8. b   9. b
10. b

Chapter 5

Key Terms Matching
1. j **2.** h **3.** p **4.** d **5.** k **6.** n **7.** l **8.** c **9.** q
10. b **11.** o **12.** i **13.** e **14.** a **15.** f **16.** m
17. g

Review Questions
18. volume, surface area **19.** previously existing cells
20. cystic fibrosis **21.** hydrophilic, hydrophobic
22. Carbohydrate **23.** bind to hormones **24.** b **25.** d
26. a. nucleus b. nucleolus c. mitochondrion
d. ribosomes e. plasma membrane **27.** T **28.** T
29. T **30.** b **31.** c **32.** i **33.** f **34.** h **35.** d
36. a **37.** g **38.** j **39.** e **40.** o **41.** n **42.** k
43. m **44.** l **45.** a. outer membrane b. inner
membrane c. matrix d. crista **46.** a. outer membrane
b. inner membrane c. granum d. thylakoid e. stroma

47. cytoskeleton **48.** actin, tubulin **49.** d **50.** a
51. e **52.** b **53.** collagen **54.** F Blood cells placed
in pure water will burst because there is a net
movement of water into the cell. **55.** F Osmosis is the
movement of water molecules across a semi-permeable
membrane which continues until equilibrium has been
reached. **56.** b **57.** Phagocytosis **58.** c
59. facilitated diffusion requires a carrier protein.
60. Active transport, ATP **61.** sodium-potassium
pump **62.** cell surface proteins

Chapter Test
1. d **2.** e **3.** a **4.** a **5.** d **6.** b **7.** d **8.** e **9.** d
10. a **11.** c **12.** a **13.** a **14.** a **15.** b **16.** b
17. e **18.** b **19.** c **20.** c **21.** a

Chapter 6

Key Terms Matching
1. e **2.** a **3.** c **4.** d **5.** b

Review Questions
6. energy, thermodynamics **7.** kilocalorie **8.** F The
quantity of energy in the universe is constant. **9.** F
The First Law of Thermodynamics concerns the
amount of energy in the universe. **10.** T
11. Catalysis, activation **12.** exergonic, endergonic

13. c **14.** c **15.** pH, temperature **16.** feedback
inhibition **17.** active **18.** T **19.** F When the
terminal phosphate group is broken off of an ATP
molecule, a significant quantity of energy is released.
20. T

Chapter Test
1. e **2.** a **3.** c **4.** e **5.** b **6.** a **7.** c **8.** b **9.** c
10. b **11.** a

Chapter 7

Key Terms Matching
1. d **2.** f **3.** e **4.** a **5.** h **6.** g **7.** b **8.** c

Review Questions
9. c **10.** photosystems **11.** photons **12.** a **13.** b
14. chemiosmosis **15.** ATP, NADPH **16.** oxygen
17. b **18.** a **19.** b **20.** a

Chapter Test
1. d **2.** c **3.** b **4.** c **5.** b **6.** d **7.** a **8.** c **9.** b
10. b

Chapter 8

Key Terms Matching
1. f **2.** b **3.** h **4.** g **5.** d **6.** a **7.** e **8.** c

Review Questions
9. oxygen, some other inorganic compound **10.**
methanogens, sulfur bacteria **11.** F The step that takes

place before respiration occurs is called glycolysis.
12. d **13.** d **14.** b **15.** c **16.** d

Chapter Test
1. d **2.** c **3.** a **4.** a **5.** b **6.** c **7.** c **8.** b **9.** d
10. a

Chapter 9

Key Terms Matching
1. f 2. i 3. j 4. c 5. g 6. a 7. e 8. d 9. h
10. b

Review Questions
11. F The term "simple cell cycle" is a reference to the production of two prokaryotic cells by binary fission.
12. T 13. e 14. d 15. karyotype, homologous pairs
16. Walther Fleming 17. diploid 18. T 19. T
20. b 21. a 22. metastases

23. chemical mutagens, viruses 24. F p53 protein monitors DNA and destroys cells with damaged DNA.
25. T 26. c 27. c

Chapter Test
1. c 2. d 3. c 4. c 5. b 6. c 7. a 8. e 9. d
10. b

Chapter 10

Key Terms Matching
1. f 2. g 3. i 4. e 5. a 6. c 7. j 8. h 9. d
10. b

Review Questions
11. c 12. b 13. d 14. a 15. meiosis produces gametes, fertilization 16. somatic 17. b 18. c 19. d 20. a 21. synapsis, reduction division 22. 23
23. 8 million

Chapter Test
1. c 2. d 3. b 4. a 5. d 6. b 7. d 8. a 9. c
10. d

Chapter 11

Key Terms Matching
1. j 2. n 3. f 4. l 5. h 6. b 7. e 8. m 9. d
10. a 11. k 12. c 13. g 14. i

Review Questions
15. b 16. e 17. genotype, phenotype 18. 1:2:1
19. heterozygous 20. recessive 21. c 22. c 23. F When doing dihybrid crosses Mendel found the F_2 generation showed a 9:3:3:1 ratio of phenotypes. 24. T
25. amino acid 26. an interaction between the products of two genes in which one of the genes modifies the phenotypic expression produced by the other. 27. b 28. F In pleiotrophy, one gene affects many traits. 29. T 30. a 31. autosomes
32. Turner syndrome, Klinefelter syndrome 33. a
34. d 35. c 36. b 37. c 38. one
39. heterozygous, malaria 40. Jews, 3,500
41. dominant allele 42. c 43. a 44. b

Chapter Test
1. e 2. c 3. b 4. a 5. c 6. c 7. b 8. d 9. e
10. b 11. c 12. b 13. c 14. d 15. b 16. d
17. b 18. e 19. a

Chapter 12

Key Terms Matching
1. d 2. b 3. l 4. m 5. a 6. k 7. f 8. e 9. i
10. g 11. h 12. j 13. c

Review Questions
14. did not, still did not 15. transformation
16. protein 17. DNA 18. T 19. F Hershey and
Chase found the hereditary material in bacteriophages

was DNA. 20. a 21. d 22. Semiconservative
replication 23. ^{15}N 24. Replication fork
25. in germ-line tissue 26. c 27. a 28. d

Chapter Test
1. a 2. c 3. b 4. c 5. d 6. e 7. b 8. a 9. b
10. d

Chapter 13

Key Terms Matching
1. j 2. a 3. k 4. e 5. b 6. d 7. h 8. l 9. f
10. c 11. i 12. m 13. n 14. g

Review Questions
15. T 16. T 17. amino acid, stop 18. universal
19. b 20. e 21. multigene families 22. transposons

23. F Activators are regulatory proteins that assist in the
unzipping of the double helix in some genes. 24. T

Chapter Test
1. d 2. c 3. b 4. c 5. c 6. c 7. a 8. b 9. e
10. b 11. d 12. c

Chapter 14

Key Terms Matching
1. j 2. a 3. e 4. h 5. i 6. k 7. l 8. c 9. g
10. f 11. d 12. b

Review Questions
13. b 14. c 15. e 16. F Noncoding DNA sequences
called introns separate shorter coding regions called
exons. 17. T 18. T 19. F–Approximately 1 – 1.5%
of the DNA in the human genome codes for functional
proteins. 20. a 21. e 22. e 23. b 24. d 25. c
26. a 27. pest resistance, herbicide resistance,

improved nutrition, hardiness 28. d 29. Mammary
30. electric shock 31. Genomic imprinting and DNA
reprogramming often do not occur properly. 32.
pluripotent 33. In the inner cell mass of the
blastocyst from embryos. 34. immunological rejection
of donor tissue by recipient must be overcome 35. T

Chapter Test
1. e 2. d 3. a 4. b 5. b 6. e 7. c 8. c 9. c
10. d

Chapter 15

Key Terms Matching
1. k 2. l 3. h 4. g 5. i 6. a 7. b 8. j 9. e
10. f 11. c 12. m 13. d

Review Questions
14. microevolutionary, reproduction
15. variants 16. a 17. d 18. T 19. T
20. Homologous, analogous 21. intelligent design
22. b 23. d 24. a 25. c 26. a 27. c 28. d 29.
b 30. disruptive 31. stabilizing 32. F People who

are heterozygous for sickle-cell anemia are less
susceptible to malaria. 33. T 34. d 35. d 36. e
37. Species are composed of populations whose
members mate and produce fertile offspring. 38. b
39. c 40. d 41. c 42. a 43. e 44. b 45. b 46.
c 47. a

Chapter Test
1. a 2. c 3. a 4. c 5. e 6. a 7. e 8. b 9. a
10. b 11. c 12. a 13. e 14. c 15. c

Chapter 16

Key Terms Matching
1. c 2. d 3. f 4. a 5. g 6. k 7. i 8. j 9. c
10. h 11. b

Review Questions
12. a. land dwellers b. air dwellers c. water dwellers
13. polynomial 14. F Common names are bound to cause confusion when discussing organisms. 15. T
16. a 17. e 18. c 19. T 20. clade 21. phylogeny
22. earlier 23. e 24. c 25. f 26. d 27. a 28. b
29. b 30. c 31. a. methanogens b. extreme thermophiles c. halophiles 32. salt, bacteriorhodopsin 33. b

Chapter Test
1. a 2. b 3. d 4. d 5. e 6. b 7. d 8. b 9. a
10. a 11. e 12. a 13. d 14. c 15. b

Chapter 17

Key Terms Matching
1. p 2. l 3. h 4. e 5. b 6. i 7. a 8. g 9. m
10. o 11. f 12. k 13. n 14. j 15. c 16. d

Review Questions
17. d 18. microspheres 19. a 20. e 21. b 22. b
23. d 24. c 25. a 26. b 27. a 28. c 29. photosynthetic bacteria 30. DNA 31. c
32. phagotrophs, holozoic feeders 33. d 34. cellular specialization 35. 15 36. sporozoans 37. photosynthetic 38. F Fungi are heterotrophs and are not plants. 39. F Molds consist of long chains of cells called hyphae and the walls separating one cell from another are septa. 40. T 41. b 42. a 43. c 44. d
45. absorb nutrients, photosynthesize 46. lignin 47. T

Chapter Test
1. b 2. c 3. e 4. d 5. c 6. b 7. b 8. a 9. c
10. d 11. e

Chapter 18

Key Terms Matching
1. k 2. q 3. n 4. o 5. l 6. m 7. r 8. a 9. d
10. i 11. j 12. c 13. f 14. h 15. b 16. p 17. g
18. e

Review Questions
19. cuticle, stomata 20. diploid, haploid
21. vascular tissue 22. d 23. F The mosses were the first plants to evolve vascular tissue. 24. F The common name for members of phylum Anthocerophyta is the hornworts. 25. e 26. d 27. a
28. microgametophyte, pollen grain 29. seed 30. a
31. a. stamen b. stigma c. style d. carpel e. ovary
f. sepal g. petal 32. yellow or blue, hummingbirds
33. wind 34. T 35. F Corn is a monocot plant.
36. animals, water, wind

Chapter Test
1. c 2. c 3. e 4. c 5. d 6. b 7. a 8. d 9. a
10. a 11. b 12. c 13. b 14. d 15. a 16. c

Chapter 19

Key Terms Matching
1. k 2. d 3. g 4. i 5. a 6. j 7. e 8. l 9. f
10. c 11. h 12. b 13. o 14. q 15. m 16. t 17. s 18. u 19. r 20. n 21. p

Review Questions
22. a 23. d 24. b 25. c 26. molecular systematics
27. e 28. d 29. cnidocytes, nematocyst 30. layers of tissue 31. d 32. a 33. a 34. between the endoderm and the mesoderm, within the mesoderm
35. b 36. b 37. a 38. b 39. segmented bodies
40. primary induction 41. evolutionary flexibility
42. c 43. d 44. cleavage 45. radially, spirally

46. T 47. T 48. notochord, nerve cord, pharyngeal slits, postanal tail 49. plants 50. ocean 51. gills, vertebral column, single-loop blood circulation, nutritional deficiencies 52. outpocketing of the pharynx 53. b 54. e 55. d 56. c 57. b 58. a
59. d 60. c 61. a 62. b

Chapter Test
1. a 2. b 3. c 4. c 5. d 6. e 7. d 8. b 9. c
10. d 11. e 12. e 13. b 14. d 15. c 16. a 17. b 18. a 19. d 20. d 21. b

Chapter 20

Key Terms Matching
1. i 2. m 3. e 4. b 5. k 6. a 7. j 8. d 9. c
10. l 11. h 12. g 13. f

Review Questions
14. c 15. b 16. d 17. c 18. aquatic ecosystems
19. primary producers 20. transpiration
21. precipitation, evaporation 22. F Carbon captured from the atmosphere by photosynthesis can be returned by respiration, combustion and erosion. 23. F Carbon dioxide in the atmosphere represents the most plentiful

source of carbon in the ecosystem. 24. plants or organisms that have consumed plants 25. e 26. T
27. T 28. c 29. Gyres 30. El Niño, a mystery
31. shallow waters 32. plankton 33. b 34. a 35. b
36. a 37. mountains, climatic effects, different sea temperatures

Chapter Test
1. e 2. c 3. a 4. d 5. c 6. a 7. b 8. c 9. e
10. d 11. b 12. b 13. c 14. c 15. d 16. d
17. f 18. a 19. g 20. d 21. c 22. e 23. b

Chapter 21

Key Terms Matching
1. o 2. j 3. d 4. p 5. a 6. c 7. e 8. m 9. k
10. h 11. b 12. f 13. n 14. g 15. l 16. i

Review Questions
17. b 18. d 19. Density-independent, population size
20. maximal sustainable yield 21. density-dependent
22. a 23. c 24. a 25. d 26. b 27. T 28. T
29. b 30. e 31. F Gause's principle states that two species cannot occupy the same area indefinitely.
32. T 33. Sympatric 34. sympatric

35. F Symbiosis, or two species living together, is often seen in nature. 36. F In commensalism one species benefits while the other neither benefits nor is harmed.
37. b 38. d 39. Parasitoids 40. brood 41. T
42. c 43. b 44. b 45. warning coloration, camouflage, chemical defenses 46. a 47. a 48. c
49. b

Chapter Test
1. d 2. c 3. b 4. c 5. e 6. c 7. c 8. a 9. b
10. b 11. a 12. b 13. c

Chapter 22

Key Terms Matching
1. g   2. e   3. a   4. f   5. h   6. b   7. d   8. c

Review Questions
9. proximate, ultimate   10. b   11. b   12. behavioral genetics   13. F Nonassociative learning does not require an animal to form an association between stimulus and response.   14. Classical conditioning involves the use of a conditioned stimulus, whereas with operant conditioning the animal associates a stimulus with reward or punishment.   15. c   16. T   17. behavioral ecology   18. Natural selection favors behaviors of food acquisition and territorial defense that maximize energy gains.   19. use of sun and stars, magnetic fields, wave action   20. c   21. Dance language communicates direction and distance, whereas Wenner and others believed flower odor was the cue.   22. kin selection   23. d   24. heredity, learning

Chapter Test
1. c   2. d   3. c   4. a   5. c   6. b   7. c

Chapter 23

Key Terms Matching
1. d   2. h   3. b   4. j   5. c   6. a   7. i   8. f   9. e   10. g

Review Questions
11. c   12. a   13. T   14. T   15. 1, 6   16. a
17. habitat loss, overpopulation, introduced species
18. F Laws and taxes designed to curb the effects of pollution will only work if they are actually implemented.   19. safety, security and waste disposal
20. the accident at Three Mile Island   21. 20, half
22. c
23. a. Habitat restoration involves restoring original communities (natural inhabitants) of plants and animals, removal of introduced species, cleanup of pollution. b. Captive propagation involves introduction of wild-caught individuals into captive breeding programs, maintaining and increasing genetic diversity, and preserving those species essential to a particular ecosystem. c. Conservation of ecosystems involves maintaining and preserving large areas or reserves that are relatively undisturbed.   24. (Answers will vary.)

Chapter Test
1. b   2. b   3. b   4. c   5. d   6. e   7. e   8. *Exxon Valdez*   9. chlorofluorocarbons   10. greenhouse
11. Recycling   12. topsoil   13. groundwater
14. 12,000 to 13,000   15. urban centers

Chapter 24

Key Terms Matching
1. h   2. e   3. k   4. m   5. p   6. c   7. q   8. a   9. l
10. n   11. b   12. o   13. f   14. i   15. d   16. g   17. j

Review Questions
18. a   19. e   20. b   21. c   22. a   23. burrowing
24. expanding   25. radial, anus   26. spiral, mouth
27. c   28. b   29. d   30. e   31. a   32. c   33. b
34. a   35. d   36. c   37. b   38. smooth, skeletal, cardiac   39. myofibrils   40. a. dendrites b. cell body c. axon d. nucleus   41. 206, 80, 126   42. c   43. b
44. a   45. actin, myosin   46. a   47. c

Chapter Test
1. a   2. d   3. d   4. b   5. c   6. e   7. b   8. d   9. b
10. c   11. b   12. c   13. d   14. e   15. a

Chapter 25

Key Terms Matching
1. f 2. r 3. g 4. l 5. j 6. a 7. d 8. o 9. m
10. h 11. k 12. b 13. e 14. q 15. i 16. p
17. c 18. n

Review Questions
19. b 20. a 21. d 22. b 23. white blood cells,
destroy bacteria and dead cells 24. movement of the
body's muscles 25. c 26. d 27. a 28. b 29. c
30. cutaneous respiration 31. systemic circulation
32. c 33. d 34. a 35. b
36. systolic, diastolic 37. listening, monitoring blood
pressure changes, measuring waves of depolarization
created by its contractions 38. a 39. c 40. b

Chapter Test
1. c 2. d 3. a 4. a 5. d 6. b 7. a 8. e 9. c
10. d 11. b 12. a 13. e 14. b 15. d 16. c

Chapter 26

Key Terms Matching
1. k 2. e 3. f 4. c 5. h 6. b 7. d 8. o 9. j
10. l 11. n 12. g 13. a 14. i 15. m

Chapter Test
1. b 2. d 3. a 4. c 5. d 6. a 7. e 8. e 9. a
10. d 11. b 12. b

Review Questions
16. a 17. d 18. T 19. F Air flows through the lungs
of birds in one direction, from back to front. 20. b
21. d 22. d 23. e 24. unchecked cell division 25.
damaged DNA 26. not smoking

Chapter 27

Key Terms Matching
1. b 2. h 3. l 4. f 5. o 6. p 7. n 8. e 9. c
10. g 11. i 12. k 13. a 14. d 15. m 16. j

Review Questions
17. b 18. a 19. c 20. e 21. d 22. c 23. a
24. b 25. intracellularly, extracellularly, digestive
cavity 26. cecum, cellulose 27. a. incisors b. canine
c. premolars d. molars 28. T 29. T 30. b 31. d
32. c 33. F Edema may indicate a drop in plasma
proteins due to a malfunctioning liver. 34. T

Chapter Test
1. c 2. e 3. a 4. c 5. e 6. c 7. d 8. b 9. c
10. b 11. d 12. e 13. b 14. a 15. c

Chapter 28

Key Terms Matching
1. k 2. p 3. a 4. f 5. l 6. j 7. n 8. g 9. c
10. m 11. h 12. b 13. e 14. i 15. d 16. o

Review Questions
17. hypothalamus 18. Reptiles 19. decreases 20.
protonephridia, flame cells
21. Malpighian tubules 22. T 23. T 24. a. nephron
b. renal cortex c. renal medulla d. renal artery
e. ureter 25. amino acids, nucleic acids 26.
mammals 27. ammonia

Chapter Test
1. c 2. d 3. e 4. c 5. b 6. b 7. d 8. b 9. a
10. c

Chapter 29

Key Terms Matching
1. m 2. h 3. e 4. o 5. k 6. b 7. f 8. c 9. n
10. a 11. i 12. p 13. d 14. g 15. j 16. l

Review Questions
17. a. epidermis b. dermis c. subcutaneous layer
d. sebaceous gland e. arrector pili muscle f. sweat
gland 18. e 19. e 20. d 21. c 22. b 23. a
24. major histocompatibility proteins
25. macrophages, MHC 26. c 27. T 28. F
Memory B cells remain in the body's tissues,

sometimes for life. 29. c 30. lectins 31. fish with
jaws 32. *nef*, slowing its replication 33. they are
similar 34. c 35. b 36. multiple sclerosis, diabetes
type 1, allergies, asthma 37. F HIV is the cause of
acquired immunodeficiency syndrome and is not highly
contagious. 38. T

Chapter Test
1. d 2. b 3. c 4. c 5. e 6. d 7. d 8. b 9. d
10. c 11. a 12. d 13. d 14. b 15. f 16. d
17. a 18. c 19. e

Chapter 30

Key Terms Matching
1. k 2. r 3. m 4. b 5. e 6. n 7. j 8. p 9. a
10. f 11. d 12. i 13. g 14. q 15. c 16. o 17. l
18. h

Review Questions
19. neurons, supporting cells 20. brain 21. b 22. a
23. T 24. T 25. e 26. blocking reabsorption of
serotonin 27. neuromodulators 28. c 29. b 30. d
31. a 32. c 33. d 34. up and down the length of the
spinal cord. 35. myelin 36. F After a meal of fried
chicken the autonomic nervous system would play a
role in stimulating the gallbladder to release bile. 37.
T 38. c 39. interoceptors, exteroceptors
40. Sensory receptors 41. T 42. e 43. b 44.

bitter, sour, salty, sweet 45. a. ear canal b. ossicles
c. semicircular canals d. auditory nerve e. cochlea
f. eustachian tube g. eardrum 46. timing how long it
takes for a sound to return 47. lateral line system
48. b 49. d 50. e 51. c 52. a 53. a. cornea
b. lens c. suspensory ligament d. sclera e. retina
f. choroid g. optic nerve 54. F The pit organ is
composed of two chambers separated by a membrane. I
detects differences in heat, which helps the pit viper
locate its prey.

Chapter Test
1. c 2. c 3. b 4. c 5. a 6. b 7. a 8. b 9. d
10. e 11. b 12. a 13. c 14. c 15. d 16. e
17. b 18. a

Chapter 31

Key Terms Matching
1. e 2. b 3. j 4. g 5. a 6. f 7. d 8. k 9. i
10. h 11. c

Review Questions
12. hormone, protein receptor 13. an adjacent cell,
through the body 14. cholesterol, four 15. in the
cytoplasm 16. a 17. a 18. a 19. c 20. d 21. b
22. F 10% of diabetics suffer from type I diabetes.

23. F In type II diabetes the number of insulin
receptors is low. 24. b 25. d 26. c

Chapter Test
1. e 2. c 3. d 4. c 5. e 6. a 7. d 8. a 9. a
10. c 11. d 12. g 13. f 14. e 15. a 16. h
17. c 18. b

Chapter 32

Key Terms Matching
1. i   2. n   3. d   4. b   5. p   6. m   7. h   8. a   9. r
10. c   11. q   12. g   13. j   14. e   15. o   16. k   17. f
18. l

Review Questions
19. d   20. a   21. b   22. c   23. a. acrosome b. nucleus c. body d. flagellum   24. a. oviducts
b. uterus c. cervix d. vagina e. ovary   25. meiosis
26. prolactin, oxytocin   27. c   28. b   29. d   30. e
31. a   32. Lanugo   33. fetal alcohol syndrome   34. d
35. a   36. c   37. b

Chapter Test
1. b   2. e   3. d   4. c   5. b   6. b   7. d   8. g   9. d
10. h   11. b   12. a   13. c   14. e   15. f

Chapter 33

Key Terms Matching
1. b   2. f   3. i   4. j   5. c   6. k   7. m   8. h   9. e
10. l   11. a   12. g   13. d

Review Questions
14. vascular cambium, cork cambium   15. to divide
16. T   17. T   18. a. cortex b. epidermis c. phloem
d. xylem e. primary xylem f. apical meristem
19. apical meristem   20. tips, in circumference

21. a. upper epidermis b. palisade mesophyll c. spongy
mesophyll d. cuticle e. guard cell f. stoma g. vein
22. e   23. b   24. water pressure   25. carbohydrates
26. potassium ions

Chapter Test
1. a   2. b   3. e   4. b   5. b   6. c   7. d   8. b   9. d
10. b   11. a   12. e   13. f   14. i   15. g   16. b
17. d   18. a   19. h   20. c

Chapter 34

Key Terms Matching
1. k   2. d   3. j   4. f   5. a   6. c   7. b   8. e   9. g
10. i   11. h

Review Questions
12. microspores   13. pollen tube   14. F In order for
pollination to occur, the pollen must be placed on the
stigma of the flower.   15. T   16. T   17. water,
oxygen   18. cotyledons   19. auxin, cytokinins,
gibberellins, ethylene, abscisic acid   20. a   21. d
22. 60, one   23. stomata, guard cells   24. e   25. b
26. Tropisms, gravity, touch, light

Chapter Test
1. c   2. b   3. a   4. c   5. d   6. e   7. c   8. a   9. c
10. d   11. e   12. a   13. d   14. c   15. b